Proteomics and Food Analysis

Proteomics and Food Analysis: Principles, Techniques, and Applications

Editor

Mónica Carrera

MDPI • Basel • Beijing • Wuhan • Barcelona • Belgrade • Manchester • Tokyo • Cluj • Tianjin

Editor
Mónica Carrera
Food Technology
Spanish National Research
Council (CSIC), Institute of
Marine Research (IIM)
Vigo
Spain

Editorial Office
MDPI
St. Alban-Anlage 66
4052 Basel, Switzerland

This is a reprint of articles from the Special Issue published online in the open access journal *Foods* (ISSN 2304-8158) (available at: www.mdpi.com/journal/foods/special_issues/Proteomics_Food_Analysis_Principles_Techniques_Applications).

For citation purposes, cite each article independently as indicated on the article page online and as indicated below:

LastName, A.A.; LastName, B.B.; LastName, C.C. Article Title. *Journal Name* **Year**, *Volume Number*, Page Range.

ISBN 978-3-0365-2441-2 (Hbk)
ISBN 978-3-0365-2440-5 (PDF)

© 2021 by the authors. Articles in this book are Open Access and distributed under the Creative Commons Attribution (CC BY) license, which allows users to download, copy and build upon published articles, as long as the author and publisher are properly credited, which ensures maximum dissemination and a wider impact of our publications.

The book as a whole is distributed by MDPI under the terms and conditions of the Creative Commons license CC BY-NC-ND.

Contents

About the Editor vii

Preface to "Proteomics and Food Analysis: Principles, Techniques, and Applications" ix

Mónica Carrera
Proteomics and Food Analysis: Principles, Techniques, and Applications
Reprinted from: *Foods* **2021**, *10*, 2538, doi:10.3390/foods10112538 1

Dahlia Daher, Barbara Deracinois, Philippe Courcoux, Alain Baniel, Sylvie Chollet, Rénato Froidevaux and Christophe Flahaut
Sensopeptidomic Kinetic Approach Combined with Decision Trees and Random Forests to Study the Bitterness during Enzymatic Hydrolysis Kinetics of Micellar Caseins
Reprinted from: *Foods* **2021**, *10*, 1312, doi:10.3390/foods10061312 5

Ana G. Abril, Mónica Carrera, Karola Böhme, Jorge Barros-Velázquez, Benito Cañas, José-Luis R. Rama, Tomás G. Villa and Pilar Calo-Mata
Proteomic Characterization of Bacteriophage Peptides from the Mastitis Producer *Staphylococcus aureus* by LC-ESI-MS/MS and the Bacteriophage Phylogenomic Analysis
Reprinted from: *Foods* **2021**, *10*, 799, doi:10.3390/foods10040799 21

Linda Monaci, Elisabetta De Angelis, Rocco Guagnano, Aristide P. Ganci, Ignazio Garaguso, Alessandro Fiocchi and Rosa Pilolli
Validation of a MS Based Proteomics Method for Milk and Egg Quantification in Cookies at the Lowest VITAL Levels: An Alternative to the Use of Precautionary Labeling
Reprinted from: *Foods* **2020**, *9*, 1489, doi:10.3390/foods9101489 43

Tae-Ho Ham, Yoonjung Lee, Soon-Wook Kwon, Myoung-Jun Jang, Youn-Jin Park and Joohyun Lee
Increasing Coverage of Proteome Identification of the Fruiting Body of *Agaricus bisporus* by Shotgun Proteomics
Reprinted from: *Foods* **2020**, *9*, 632, doi:10.3390/foods9050632 57

Robert Stryiński, Elżbieta Łopieńska-Biernat and Mónica Carrera
Proteomic Insights into the Biology of the Most Important Foodborne Parasites in Europe
Reprinted from: *Foods* **2020**, *9*, 1403, doi:10.3390/foods9101403 69

Mónica Carrera, Carmen Piñeiro and Iciar Martinez
Proteomic Strategies to Evaluate the Impact of Farming Conditions on Food Quality and Safety in Aquaculture Products
Reprinted from: *Foods* **2020**, *9*, 1050, doi:10.3390/foods9081050 103

About the Editor

Mónica Carrera

Dr. Mónica Carrera is a tenured scientist at the Institute of Marine Research (IIM), Spanish National Research Council (CSIC) in Vigo (2020). She received her PhD in Biology with Honors at IIM-CSIC, 2008, in the field of proteomics and seafood authenticity and quality under the supervision of Prof. Dr. José M. Gallardo, Dr. Carmen Piñeiro and Dr. Benito Cañas. From 2008 to 2013, she was working in the Proteomics group headed by Prof. Dr. Ruedi Aebersold at the Institute of Molecular Systems Biology (IMSB; ETH Zürich) in the field of proteomics and systems biology. In 2015, as a Proteomics Senior Researcher, Dr. Carrera was invited by the company Thermo Fisher Scientific (CA, USA) to work in collaboration with Dr. Daniel Lopez-Ferrer in the field of proteomics and food safety. Dr. Carrera is an internationally recognized expert on advanced proteomics in the context of seafood quality and safety. She has been recognized by 10 scientific awards (best PhD Thesis 2009; best Publication 2010, 2016, 2019; best Congress Chairman 2012; honorific distinction CSIC 2017; Congress Prize 2019, 2021; Excellent European Female Researcher 2021).

Preface to "Proteomics and Food Analysis: Principles, Techniques, and Applications"

This book, *Proteomics and Food Analysis: Principles, Techniques, and Applications,* edited by Dr. Mónica Carrera and published by MDPI, is an excellent collection of a wide range of proteomics approaches applied in food analysis.

Proteomics methodologies are an advantageous strategy for food science studies, where research institutions, agencies, food industries, and regulatory laboratories are combining efforts to acquire necessary knowledge on food composition, quality and safety.

The potential of proteomics in food analysis is highlighted in this book, which contains one editorial article and six scientific manuscripts covering different applications of proteomics methodologies to ensure food quality and safety. This book is an ideal and up-to-date guide for researchers seeking to understand the proteomics methodologies applied to different foods.

Finally, the editor wants to express her gratitude to all the coauthors for their assistance in the preparation of this book.

Mónica Carrera
Editor

foods

Editorial

Proteomics and Food Analysis: Principles, Techniques, and Applications

Mónica Carrera

Food Technology Department, Institute of Marine Research (IIM), Spanish National Research Council (CSIC), 36208 Vigo, Pontevedra, Spain; mcarrera@iim.csic.es; Tel.: +34-986231930

Proteomics can be considered the discipline of the large-scale analysis of proteins in a particular biological system. Proteomics comprises the study of the structure and function of proteins, the quantification of protein abundance, the determination of protein intracellular location, the analysis of protein modifications and the study of protein–protein interaction networks. Mass spectrometry (MS)-based proteomics has been recognized as an indispensable tool to precisely identify and quantify thousands of proteins from complex protein samples, and is used in the majority of proteomics studies. Moreover, bioinformatics treatment of MS data has increased the scale of proteomics tools, representing a powerful strategy for high-throughput protein and peptide identification and quantification. In this regard, two sequential food proteomics approaches (discovery food proteomics and targeted food proteomics) are the main proteomic tools used for food quality and safety studies. The aim of discovery food proteomics is to analyze a particular proteome to identify potential food protein/peptide biomarkers, commonly using a bottom-up proteomics strategy in which the proteins of interest are digested into peptides using proteases (i.e., trypsin), and the resulting peptides are analyzed by MS. Then, targeted food proteomics methods are used to search for the peptide biomarkers selected in the discovery phase in biological samples or in food products with high precision, sensitivity and reproducibility. Thus, proteomics methodologies are an advantageous strategy for food science studies, where research institutions, agencies, food industries, and regulatory laboratories are combining efforts to acquire necessary knowledge on food composition, quality, and safety. The potential of proteomics in food analysis is highlighted in this special issue on different subjects concerning food quality and safety. The six scientific papers that contribute to this Special Issue, "Proteomics and Food Analysis: Principles, Techniques, and Applications", provide an excellent overview of the wide-ranging proteomics approaches applied to food analysis.

In this context, the manuscript published by Daher et al., "Sensopeptidomic kinetic approach combined to decision trees and random forest to study the bitterness during enzymatic hydrolysis kinetics of micellar caseins", reports the proteomic characterization of protein hydrolysates responsible for the unpleasant taste in milk [1]. Milk protein hydrolysates have significant advantages in sport nutrition and elderly and infant nutrition, as the use of these hydrolysates induces a very rapid release of amino acids in the blood, which maximizes muscle protein anabolism and facilitates body recovery and nutrition. However, hydrolyzed proteins can sometimes have an unpleasant bitter taste or off flavors, which limits the breadth of their applications in nutrition. In this article, the authors performed an interesting investigation of peptide characterization of micellar caseins using proteolytic enzymes and liquid chromatography-electrospray-quadrupole-time-of-flight-tandem mass spectrometry (LC-ESI-Q-TOF-MS/MS) analysis. Moreover, they correlated the amino acid structure of the identified peptides with the bitterness properties of micellar casein hydrolysates by using different statistical and bioinformatics tools based on differential expression analysis, heat maps, regression trees and random forests. The authors concluded that the formulated hydrolysates may be used in the development of future food formulations such as new peptide-fortified ready-to-drink infant formulas.

An interesting proteomics approach reported by Abril et al. in "Proteomic characterization of bacteriophage peptides from the mastitis producer *Staphylococcus aureus* by LC-ESI-MS/MS and bacteriophage phylogenomic analysis" describes the characterization of bacteriophage peptides from the mastitis-causing bacterium *Staphylococcus aureus* (*S. aureus*) isolated from dairy products [2]. *S. aureus* is considered one of the major foodborne pathogens that can cause serious food intoxication in humans due to its production of endotoxins. This bacterium remains a major problem in the dairy industry due to its persistence in cows, its pathogenicity, its contagiousness and its ability to easily colonize the skin and mucosal epithelia. The authors used a shotgun proteomics approach (in a gel-free strategy where a complex mixture of food proteins is digested in solution with a protease such as trypsin, and the resulting mixture of peptides is then analyzed by LC-MS/MS) for the characterization of 20 different *S. aureus* strains. For this purpose, they utilized an LC-ESI-MS/MS-based workflow using an LTQ-Orbitrap instrument to identify relevant phage-specific peptides of several *S. aureus* strains to identify both phages and bacterial strains. Additionally, phylogenomic trees were developed to demonstrate a link between phage phylogeny and their ability to infect the same bacterial species. The authors concluded based on the data obtained for the different models of mastitis that the phage therapy using bacteriophages in this study may be considered an innovative alternative to antibiotics for the treatment of mastitis caused by *S. aureus*.

The manuscript published by Monaci et al., "Validation of a MS-based proteomics method for milk and egg quantification in cookies at the lowest VITAL levels: an alternative to the use of precautionary labeling", presents for the first time the development of a targeted proteomics approach based on the multiple reaction monitoring (MRM) method on a triple quadrupole MS instrument for milk and egg identification and quantification in processed foods such as cookies [3]. The method allows the detection of milk and egg proteins at levels lower than the 0.2 mg recommended by the Voluntary Incidental Trace Allergen Labeling (VITAL) program as necessary to help food producers avoid cross contamination. As stated by the authors, this could be used as an alternative method for precautionary allergen labeling, alongside other common detection methodologies based on proteins or nucleic acids. The authors conclude that this targeted proteomics method could represent, a promising tool to be implemented along the food chain to detect even tiny amounts of allergens contaminating food commodities.

In the article written by Ham et al., "Increasing coverage of proteome identification of the fruiting body of *Agaricus bisporus* by shotgun proteomics", the authors provide an overview on the protein identification of the fruit body of *Agaricus bisporus*, analyzing the crude protein fraction of the fruit body using a shotgun proteomics approach using multidimensional protein identification technology (MudPIT) by LC-MS/MS in an LTQ mass spectrometer [4]. Their protocol involved biphasic column separation prepared with reversed-phase resins followed by strong cation exchange material. The relative quantification of the identified proteins revealed several protein signatures that are highly abundant in the fruiting body. Functional classifications of the identified proteins were also provided by bioinformatics tools.

The manuscript produced by Stryinski et al., "Proteomic insights into the biology of the most important foodborne parasites in Europe", provides an excellent overview of the applications of proteomic methods in studies on foodborne parasites and their potential use in targeted diagnostics [5]. Discovery proteomics methods were described for the characterization and selection of protein biomarkers for selected foodborne parasites such as waterborne parasitic species (e.g., *Cryptosporidium* spp., *Giardia lamblia*, *Entamoeba histolytica*, etc.), soil- and plant-borne parasitic species (e.g., *Echinococcus multilocularis Toxocara* spp., *Ascaris* spp., *Fasciola* spp., *Trypanosoma cruzi*, etc.), meat-borne parasitic species (i.e., *Toxoplasma gondii*, *Trichinella* spp., *Taenia* spp., and *Sarcocystis* spp.) and seafood borne parasitic species (i.e., Opisthorchiidae, *Angiostrongylus cantonensis*, *Diphyllobothrium* spp., *Paragonimus* spp., and Heterophyidae). Targeted proteomics methods were also described to search for peptide biomarkers selected in the discovery phase with high precision, sensitivity and reproducibility. This paper was mainly focused on the description of targeted proteomic methods proposed for Anisakidae detection in food products.

Finally, the article published by Carrera et al., "Proteomic strategies to evaluate the impact of farming conditions on food quality and safety in aquaculture products", presents different proteomic strategies (discovery and targeted proteomics) to evaluate the impact of farming conditions on food quality and safety in aquaculture products [6]. Food quality, dietary management, fish welfare, stress response, food safety (biotic and abiotic hazards) and antibiotic resistance are the main proteomic techniques and strategies that are successfully covered in this review. The authors conclude by outlining future directions and potential perspectives, as the development and practical implementation of new advances based on protein microfluidics, protein biosensors and device digitalization offer promising research areas for the aquaculture industry and food authorities.

The "Proteomics and Food Analysis: Principles, Techniques, and Applications" Special Issue is an ideal and timely guide for researchers seeking to understand the proteome of any food biological sample. Finally, the Guest Editor, Dr. Carrera, wishes to express her gratitude to all the authors for their contribution in the preparation of this Special Issue.

Funding: This work was funded by the GAIN-Xunta de Galicia Project (IN607D 2017/01) and the Spanish AEI/EU-FEDER PID2019-103845RB-C21 project.

Institutional Review Board Statement: No applicable.

Informed Consent Statement: No applicable.

Data Availability Statement: No applicable.

Conflicts of Interest: The authors declare no conflict of interest.

References

1. Daher, D.; Deracinois, B.; Courcoux, P.; Baniel, A.; Chollet, S.; Froidevaux, R.; Flahaut, C. Sensopeptidomic kinetic approach combined with decision trees and random forests to study the bitterness during enzymatic hydrolysis kinetics of micellar caseins. *Foods* **2021**, *10*, 1312. [CrossRef] [PubMed]
2. Abril, A.G.; Carrera, M.; Böhme, K.; Barros-Velázquez, J.; Cañas, B.; Rama, J.L.R.; Villa, T.G.; Calo-Mata, P. Proteomic characterization of bacteriophage peptides from the mastitis producer *Staphylococcus aureus* by LC-ESI-MS/MS and the bacteriophage phylogenomic analysis. *Foods* **2021**, *10*, 799. [CrossRef] [PubMed]
3. Monaci, L.; De Angelis, E.; Guagnano, R.; Ganci, A.P.; Garaguso, I.; Fiocchi, A.; Pilolli, R. Validation of a MS based proteomics method for milk and egg quantification in cookies at the lowest VITAL levels: An alternative to the use of precautionary labeling. *Foods* **2020**, *9*, 1489. [CrossRef] [PubMed]
4. Ham, T.H.; Lee, Y.; Kwon, S.W.; Jang, M.J.; Park, Y.J.; Lee, J. Increasing coverage of proteome identification of the fruiting body of *Agaricus bisporus* by shotgun proteomics. *Foods* **2020**, *9*, 632. [CrossRef] [PubMed]
5. Stryiński, R.; Łopieńska-Biernat, E.; Carrera, M. Proteomic insights into the biology of the most important foodborne parasites in Europe. *Foods* **2020**, *9*, 1403. [CrossRef] [PubMed]
6. Carrera, M.; Piñeiro, C.; Martinez, I. Proteomic strategies to evaluate the impact of farming conditions on food quality and safety in aquaculture products. *Foods* **2020**, *9*, 1050. [CrossRef] [PubMed]

Peptidomic Kinetic Approach Combined with Decision Trees and Random Forests to Study the Bitterness during Enzymatic Hydrolysis Kinetics of Micellar Caseins

Dahlia Daher [1,2], Barbara Deracinois [1], Philippe Courcoux [3,4], Alain Baniel [2], Sylvie Chollet [1], Renato Froidevaux [1] and Christophe Flahaut [1,*]

[1] UMR Transfrontalière 1158 BioEcoAgro, Univ. Lille, INRAe, Univ. Liège, UPJV, JUNIA, Univ. Artois, Univ. Littoral Côte d'Opale, ICV—Institut Charles Viollette, 59000 Lille, France; d.daher@ingredia.com (D.D.); barbara.deracinois@univ-lille.fr (B.D.); sylvie.chollet@junia.com (S.C.); renato.froidevaux@univ-lille.fr (R.F.)
[2] Ingredia S.A. 51 Av. Lobbedez-CS 60946, CEDEX, 62033 Arras, France; a.baniel@ingredia.com
[3] Oniris, StatSC, rue de la Géraudière, 44322 Nantes, France; philippe.courcoux@oniris-nantes.fr
[4] INRA USC1381, 44322 Nantes, France
* Correspondence: christophe.flahaut@univ-artois.fr; Tel.: +33-321791780

Abstract: Protein hydrolysates are, in general, mixtures of amino acids and small peptides able to supply the body with the constituent elements of proteins in a directly assimilable form. They are therefore characterised as products with high nutritional value. However, hydrolysed proteins display an unpleasant bitter taste and possible off-flavours which limit the field of their nutrition applications. The successful identification and characterisation of bitter protein hydrolysates and, more precisely, the peptides responsible for this unpleasant taste are essential for nutritional research. Due to the large number of peptides generated during hydrolysis, there is an urgent need to develop methods in order to rapidly characterise the bitterness of protein hydrolysates. In this article, two enzymatic hydrolysis kinetics of micellar milk caseins were performed for 9 h. For both kinetics, the optimal time to obtain a hydrolysate with appreciable organoleptic qualities is 5 h. Then, the influence of the presence or absence of peptides and their intensity over time compared to the different sensory characteristics of hydrolysates was studied using heat maps, random forests and regression trees. A total of 22 peptides formed during the enzymatic proteolysis of micellar caseins and influencing the bitterness the most were identified. These methods represent simple and efficient tools to identify the peptides susceptibly responsible for bitterness intensity and predict the main sensory feature of micellar casein enzymatic hydrolysates.

Keywords: bitterness; enzymatic hydrolysis; micellar caseins; off-flavours; peptidomics; random forests; regression trees; sensory analysis

1. Introduction

The enzymatic hydrolysis of milk proteins displays a generally unpleasant bitter taste. The perception of bitter taste plays a crucial role in their use in various application fields. Indeed, the bitter flavour of extensively hydrolysed proteins has been and continues to be a major hindrance for their use. In addition, bitterness is sometimes combined with off-flavours that also appear during hydrolysis. However, these milk protein hydrolysates have significant advantages such as in sport nutrition where the use of these hydrolysates induces a very rapid release of amino acids in the blood, which may maximise muscle protein anabolism and facilitate recovery [1]. Moreover, these hydrolysates make it possible to boost muscle synthesis in sensitive subjects such as the elderly [2]. They are also used in clinical and infant nutrition where milk protein hydrolysates are recommended for a rapid supply of amino acids while ensuring low protein allergenicity. Indeed, the allergenicity of a protein is reduced or eliminated when the protein is hydrolysed into a low molecular weight peptide composition. Moreover, milk protein hydrolysates cater to the nutritional requirements of infants and toddlers, improving milk protein digestibility and reducing frequent spit-up.

For many years, scientists have performed important studies on explaining the appearance of bitterness in hydrolysates. For example, Murray and Baker were the first authors interested in the taste of protein enzymatic hydrolysates [3]. They found a bitter taste in enzymatic hydrolysates from caseins and lactalbumin, obtained with commercial proteinases, and a neutral taste in hydrolysates obtained from gelatine. Ichikawa et al. hydrolysed caseins, soy protein, ovalbumin and gluten with a proteinase from *Bacillus subtilis* and reported the development of a pronounced bitter taste with casein hydrolysates [4]. Various factors can influence the appearance of undesirable flavours in protein enzymatic hydrolysates, such as the nature of protein substrate(s) and enzyme(s), the hydrolysis duration, the selected pH and the temperature conditions. Concerning the casein proteins, β-, αS1- and κ-caseins produce the most bitter hydrolysates [5]. The causes for the bitterness were identified as early as 1970 by Fujimaki et al. and Matoba et al. [6,7] as being the presence of specific peptides rather than free amino acids in the protein hydrolysate. For example, the free forms of L-leucine and L-phenylalanine residues are bitter, with thresholds of 15–20 mM, but Leu-Leu or Ile-Leu and Leu-Phe are more than 10 times more bitter. Kim and Li-Chan (2006) and Iwaniak et al. (2018) confirmed that the bitter taste of peptides is determined by the presence of amino acids with high hydrophobicity [8,9]. According to Iwaniak's data, the bitterness of peptides results from the presence of residues with bulky and branched side chains such as Leu, Ile, Val, Tyr, Phe and Trp. The bitterness of peptides also increases as the number of amino acids increases. Moreover, some structural characteristics, such as the diastereoisomer of the L series, the presence of a proline residue at the geometric centre and/or close to a basic amino acid, hydrophobic amino acids at N and C-terminal positions in the peptide, and two and three residues of Leu, Tyr, Phe at the C-terminal of the peptide, influence the bitterness. In addition, it has been claimed that there are no bitter peptides for lengths greater than 25 residues [7].

In a previous study [10], the comparison between the sensory characteristics and the principal components of the principal component analysis (PCA) of mass spectrometry data reveals that peptidomics constitutes a convenient, valuable, fast and economic intermediate method to evaluate the bitterness of enzymatic hydrolysates as a trained sensory panel can conduct it. Nevertheless, to go further in the understanding of the peptide-related bitterness appearance/disappearance during the hydrolysis time, an enzymatic hydrolysis kinetic study gathering a sensory evaluation and a peptidomics approach combined with machine learning algorithms were carried out. Herein, we have studied the enzymatic hydrolysis kinetics of micellar caseins subjected to hydrolyses using commercially available and food-grade proteases, allowing the production of more or less bitter hydrolysates. Organoleptic characteristics, and more particularly the bitterness, were quantified for each sample collected during the kinetics using a trained sensory panel. Then, peptides generated during hydrolysis were characterised by a peptidomics approach combining the peptide chromatographic separation by reversed-phase high-pressure liquid chromatography (RP-HPLC), the detection and fragmentation of peptides by tandem mass spectrometry (MS/MS) and the mass data management. Finally, we studied the nature of the generated peptides and their influence in the appearance of bitterness during the hydrolysis process by using a method based on differential expression analysis, heat maps, regression trees and random forests.

2. Material and Methods

2.1. Enzymatic Hydrolysis Kinetics

Micellar caseins (ratio micellar caseins/whey proteins (92:8)) were prepared by the Ingredia S.A. manufacturer (St-Pol-Sur-Ternoise, France) using industrial processes. These proteins were hydrolysed with the food grade enzymes (Table 1) Flavourzyme and Protamex that were obtained from Novozymes (Bagsvaerd, Denmark) and allowed the preparation of a kinetics named 109, and Promod 523MDP ™ and FlavorPro 937 ™ were obtained from Biocatalysts and allowed the preparation of a kinetics named 125 (Wales, UK).

Table 1. Characteristics of Novozymes proteases.

Proteases	Description	Activity *	Origin	Optimum pH	Optimum Temperature (°C)
Flavourzyme	exoprotease (aminopeptidase)/endoprotease complex	1100 LAPU/g	*Aspergillus oryzae*	5.5–7.5	50–55
Protamex	endoprotease (subtilisin)/serine endoprotease	1.5 AU-N/g	*Bacillus licheniformis* *Bacillus amyloliquefaciens*	7.0–8.0	50
Promod 523MDP ™	endoprotease complex	1200 Bromelain GDU/g	*Ananas comosus*	5.0–7.0	45–55
Flavorpro 937MDP ™	exoprotease (leucine aminopeptidase)/endoprotease complex	350 U/g	*Aspergillus oryzae*	5.0–7.0	50

* Leucine amino peptidase units per gram (LAPU/g); Anson unit per gram (AU-N/g); Gelatin digestion units per gram (GDU/g).

The enzymatic hydrolyses were performed for nine hours using a confidential recipe. Overall, the protein solution of micellar caseins (92%) was diluted with distilled water to a concentration of 10% of total nitrogenous matter and brought to the desired pH by adding NaOH (4N). The necessary enzyme quantity was then added directly if it was in liquid form or solubilised in distilled water if it was in powder form. The hydrolysis monitoring was carried out by collecting data from pH, temperature and osmometry. Then, the degree of hydrolysis (DH) was determined using Nielsen et al.'s method based on the reaction of primary amino groups with ortho-phthaldialdehyde (OPA) [11], and the DH was calculated as previously described [10].

Samples were taken every hour and the enzymes were inactivated by heating at 98 °C for 3 min. About 1.5 L of hydrolysates was dried by atomisation using the Mini Spray Dryer B-290 from BUCHI (Rungis, France). The drying process was performed following the same procedure described previously [10]. Each hour, an aliquot of each hydrolysis was frozen at −20 °C before further analyses.

2.2. Analyses of Samples
2.2.1. Sensory Analysis
Panel Composition and Training

The sensory analysis was carried out with a total of 19 healthy adults (12 females and 7 males, aged from 45 to 65 years old). They were enrolled in a training program, for 20 months, designed to identify and quantify the different descriptors chosen to characterise the hydrolysates. The descriptors were: (i) five odours (milk, fermented milk, rancid, soymilk and smelly), (ii) eleven flavours (bitter, sour, milk, sweet, mild, cheese, vanilla, salty, rancid, barn and whey) and (iii) five persistence flavours (bitter, sour, milk, sweet and cheese). The quantification of each descriptor was performed using a scale from 1 (low) to 7 (high). In this study, only bitterness data will be processed. Before starting this experiment, the performance of the assessors in terms of discrimination, repeatability and agreement was validated.

Tasting Conditions

The assessors evaluated the nine samples in a duplicate manner during four sessions (two sessions with four samples and two with five), for both kinetics. Those sessions were performed under standard sensory conditions (ISO 13299, 2003). Samples were presented in a sequential monadic way and their presentation order was based on a Williams' Latin-square arrangement. Samples were dissolved in mineral water at a concentration of 10% of dry matter and 20 mL was presented in white plastic tumblers to each assessor and served at room temperature. The tests were performed in individual booths under white lighting and at 20 ± 2 °C. No time restriction was imposed on the assessors to perform this test.

Sensory Data Analyses

For both hydrolysis kinetics (109 and 125), sensory data were first assessed by a two-way ANOVA considering the samples and the consumers as factors and the bitterness scores as the dependent variable. A Duncan's multiple range test ($p \leq 0.05$) was performed to compare the samples two by two. These statistical analyses were computed using XLStat (XLStat 2020 1.1, Paris, France).

2.2.2. Mass Spectrometry: Sample Preparation and Peptide Characterisation Using HPLC-ESI-Qtof-MS/MS and Bioinformatics Treatment

The samples were prepared and analyzed in duplicate with the same method used in a previous study [10]. Briefly, peptides were purified and concentrated using a C18 solid phase extraction and 10 µL was separated using a reversed-phase high-performance liquid chromatography and an apolar gradient of 60 min: 1% ACN/0.1% of formic acid (FA) (v/v) for 3 min, then 1 to 30% ACN/0.1% FA (v/v) for 42 min, 30 to 95% ACN/0.1% FA (v/v) for 10 min and finally 95 to 99% ACN/0.1% FA (v/v) for 5 min. The analysis of eluted peptides was performed with a Synapt-G2-Si (Waters) mass spectrometer in sensitivity, positive and data-dependent analysis (DDA) modes (HPLC-MS/MS). Several quality control (QC) samples corresponding to (i) the mixture in equivalent volume of all C18-purified samples of both kinetics, (ii) the mixture of samples of kinetics 109 and (iii) those of kinetics 125 were also analysed at the beginning, middle and end of the HPLC-MS/MS analysis session.

Raw data from all HPLC-MS/MS runs were imported in Progenesis QI for proteomics software (Version 4.1, Nonlinear Dynamics, Newcastle upon Tyne, UK). First, data filtering was conducted before peak picking where a maximum charge of +4, a retention time defined between 5 and 50 min and a minimum intensity of 1000 were applied. Then, data alignment was automatically managed by Progenesis software using one of all QC runs as reference. Subsequently, manual alignment was performed if necessary, to optimise run alignment, and data normalisation was automatically performed for principal component analysis (PCA). The filtering criteria used for the statistical comparison of mass signals of HPLC-MS/MS runs were set as follows: (i) a maximum coefficient of ANOVA less or equal to 10^{-10} and (ii) only the identified peptides. Concomitantly, Progenesis software reported the quantitative evolution of peptides in terms of normalised abundance in the different hydrolysates. The variables used are derived from the comparison of peptide maps, i.e., the position of the isotopic massifs and their intensity. The reprocessing of mass spectrometry data and database searches to identify the peptides were performed via Peaks Studio version 10+ (Bioinformatics Solutions Inc., Waterloo, ON, Canada) using the UniProt database (10 September 2018) restricted to the complete proteome of *Bos taurus* organism. The parameters of mass tolerance thresholds, number of missing cleavage sites tolerated, choice of enzyme, and false discovery rate (FDR) were the same as previously [10].

2.3. Relationship between Sensory and Mass Data

The link between identified peptides and sensory perception of the samples was investigated using various methods: a heat map with differential expression analysis, and regression trees and random forest methodologies.

2.3.1. Heat Map

A heat map was drawn from the MS-data corresponding to identified peptides from micellar caseins and their quantification in the 18 samples of both kinetics 109 and 125; the peptides corresponding to the features and the hydrolysis samples to the individuals. The heat map reflects the matrix data so that the values (normalised peptide abundance) are replaced by colour intensities ranging from yellow (low abundance) to red (high abundance). Cluster analysis was also carried out based on the heat map and the results were drawn as tree maps in the heat map [12].

Differential expression was also used to identify peptides that significantly influence the bitterness of hydrolysates. This latter was performed by merging kinetics 109 and 125. For the differential expression test, two groups were established: the group named "1" will be considered as more bitter and the "2" as less bitter. The split between these two groups was determined visually as follows: the samples were ranked in descending order of bitterness and the split was defined at the point where the greatest difference between two successive values was observed.

2.3.2. Regression Trees and Random Forest

Regression trees (RTs) optimally subdivide the samples by a set of decision rules. These rules are constructed by iteratively separating the dataset with binary splits based on the choice of one predictor variable and an associated threshold value. The random forest (RF) algorithm generates multiple trees without pruning, improving the stability of the model. This is achieved by a double process of randomisation: (i) a random selection of the predictors at each node of each tree and (ii) each tree is grown on a different random data subset, selected by bootstrapping, i.e., sampling from the initial samples with replacement. The data portion used for the training phase is known as the "in-bag" data, whereas the rest is called the "out-of-bag" data. The latter will provide estimates of predicting errors [13]: the root mean square error of this predicted value is computed on the out-of-bag samples ($RMSE_{OOB}$).

In our case, RFs consist of modelling a sensory variable (bitterness descriptor) as a function of a number of predictors (presence or absence of peptides as well as their normalised abundance). The variables correspond to the 116 identified peptides. The aim here is to find out which peptides have a strong importance in understanding the intensity of the bitter descriptor. For this purpose, 50 forests of 5000 trees have been built. RFs will allow us to obtain the importance of each peptide and a confidence interval is computed around the importance of the peptides. All the peptides whose lower bounds of the confidence interval are greater than 0 are selected and are thus involved to predict the intensity of the studied attribute.

RTs and RFs have been carried out using language R 3.5.1 [14] and the R packages rpart [15] and random Forest [16].

3. Results

3.1. Influence of Hydrolysis Kinetics on the Sensory Characteristics of Hydrolysates

Figure 1 shows the evolution of bitterness intensity and DH during kinetics 109 (a) and 125 (b). ANOVA shows that samples are significantly discriminated ($p \leq 0.05$) for both kinetics.

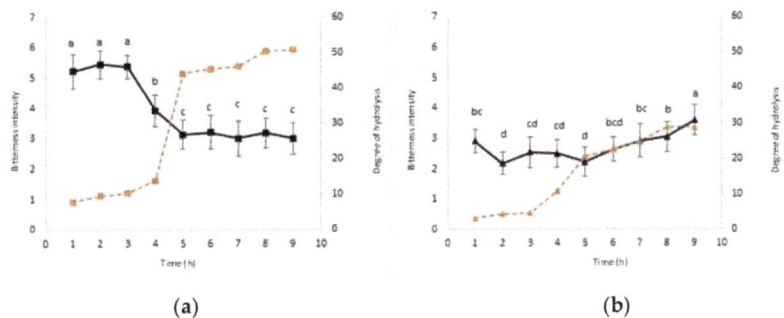

(a) (b)

Figure 1. Evolution of bitterness and DH during hydrolysis kinetics 109 (a) and 125 (b). The black kinetics 109 (■) and kinetics 125 (▲), respectively. The dotted orange lines represent the evolution of DH for kinetics 109 (■) and kinetics 125 (▲), respectively. The bitterness intensity values from 1 (low) to 7 (high) are means +/− standard deviation ($n = 2$): the different letters (a, b, c) indicate means that significantly differ among the nine samples at $p \leq 0.05$ according to Duncan's multiple range test.

Globally, the bitterness intensity (black line, Figure 1a) of the samples of the hydrolysis 109 decreases over the time. During the first three hours, the intensity is at its highest level and stagnates at the value of 5.20. Then, a decrease begins from 5.20 to 3.00 in 2 h of hydrolysis (between 3 and 5 h of hydrolysis) and remains stable at the bitterness level of 3 for the kinetics' remaining time. After nine hours of hydrolysis, the DH (dotted orange line, Figure 1a) reaches a value of 50.8%. A consequent increase in the DH is observed between the 4th (13.9%) and 5th hour (44.1%), which is concomitant with the decrease in the bitterness intensity. A Pearson correlation coefficient ($r = -0.933$; $p \leq 0.001$) between DH and bitterness score suggests that the higher the DH, the less bitterness in the samples.

Concerning kinetics 125 (black line, Figure 1b), the bitterness appears to be stable over time. The minimum bitterness value is observed after two hours of hydrolysis with an intensity of 2.18 ± 0.37. A progressive increase in bitterness is observed after the 5th hour of hydrolysis and until the end of the hydrolysis, as indicated by the bitterness values which increase from 2.23 ± 0.46 to 3.59 ± 0.50. The DH (dotted orange line, Figure 1b) increases over time to reach the maximum value of 28.9% after nine hours of hydrolysis. Here, again a high increase, ranging from 10.9% to 22.6%, is observed between the 3rd and 5th hour of hydrolysis. However, contrary to kinetics 109, the bitterness intensity increase follows the DH increase, especially from the fifth hour of kinetics.

3.2. Peptide Characterisation and Peptide Abundance Evolution during the Hydrolysis

The HPLC-MS/MS raw data obtained for the 36 withdrawn samples (nine collected samples × two kinetics × two replicates) and the nine QCs (QC 109 × three replicates, QC 125 × three replicates and QC109–125 × three replicates) were imported in Progenesis QI for proteomics software. Among the 2635 peak picked mass signals, 479 mass signals have an ANOVA < 10^{-10}, and among them, 116 mass signals were identified as milk protein peptides (Supplemental Table S1). These latter represent the global diversity, all hydrolysates combined, of identified peptides. Overall, 75 peptides from β-casein, 19 from α-S1 casein, 10 from α-S2 casein, 9 from kappa-casein, and 3 from β-lactoglobulin and no peptides from α-lactalbumin were identified. Between 114 and 116 peptides were identified per sample collected during the kinetics. The size features of identified peptides are: (i) a length comprising between 6 and 23 amino acids with a length mean of 11 amino acids and (ii) a molecular mass mean of 1303.37 ± 335.46 Da.

Peptides identified from β-casein corresponded mainly to three protein regions: Y75-G109, A116-F134 and T142-V224. The α-S1 casein- and α-S2 casein-peptides corresponded to three protein regions (G25-G48, L114-M138 and P192-P212) and two protein regions (L111-N130 and R185-A204), respectively. The κ-casein- and β-lactoglobulin-peptides corresponded to two protein regions (F39-G60 and F76-L95) and one protein region (V57-L73), respectively.

PCA was performed using the 116 identified peptides (shown as light grey numbers in Figure 2) whose amino acid sequences are gathered in Supplemental Data S1. Figure 2 shows the first two principal components and illustrates the correlations between the 36 withdrawn samples. These principal components #1 and #2 explain 82.16% of the variance and in such PCA, the more distant the groups, the more different in terms of peptide population. In the biplot presented in Figure 2, the technical replicates (same colour points) of each sample (including QCs) are close to each other, as can be seen with the examples shown with a red arrow on the PCA, indicating good technical repeatability. Moreover, the QCs of kinetics 109 (in yellow) are found at almost equal distance between the two groups formed by kinetics 109, and it is the same for those of kinetics 125 (in dark to light blue) and the QC of kinetics 109–125, which are found between the three groups represented on the PCA.

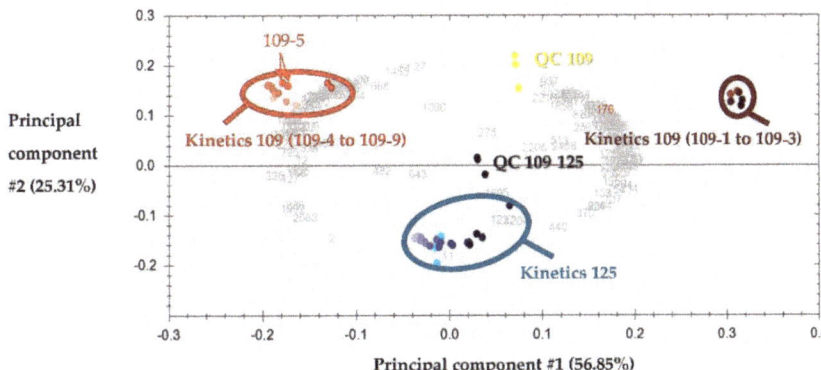

Figure 2. Principal component analysis corresponding to the 116 identified peptides of kinetics 109 and 125. In yellow the quality controls, corresponding to the equimolar mixture of the samples of kinetics 109 (QC 109), in blue "water green" those of kinetics 125 (QC 125) and in black those of all the samples combined (QC 109–125). Each QC appears as three replicates corresponding to an injection at the beginning, middle and end of the LC-MS/MS analysis. Each sample of both kinetics was analysed in replicates (samples with the same colour on the PCA as shown for the sample 109-5 (red arrows)), corresponding to a total of 36 samples: nine samples of hydrolysis kinetics 125 ranging from dark blue (125-9) to light blue (125-1) and nine samples of hydrolysis kinetics 109 ranging from brown (109-9) to very light red (109-1).

An agglomerative hierarchical clustering (AHC) allowed us to display three groups on the PCA: (i) a group circled in brown (top right) gathering samples 109-1, -2, -3, (ii) a group circled in burgundy red (top left) gathering samples 109-4, -5, -6, -7, -8, -9, and (iii) a central group circled in blue gathering all samples of kinetics 125. Notably, the evolution, according to the hydrolysis time, of peptide heterogeneity is clearly evidenced on the PCA of T1 to T9 of kinetics 125, which moves from right to left (from dark blue to light blue). As for the sensory analysis, a difference is observed between samples 109-1, -2, -3, which are significantly more bitter than the other samples of the kinetics. The samples of kinetics 125 are positioned between the two groups of kinetics 109 and thus appear to have peptide sequences common to both groups and with intermediate normalised abundances.

The Progenesis QI software uses the peptide identities and their MS-based abundance data to generate an explicit picture of the evolution of normalised abundance of the peptide during hydrolysis kinetics (Figure 3a,b). As illustrated in Figure 3a,b, the peptides FVAPFPE (αS1-CN (39–45)) and LYQEPVLGPVRGPFPI (β-CN (207–222)) are more abundant during the first three hours of kinetics 109 (left part of curves), and conversely have negligible normalised abundance in the other samples collected during kinetics 109 and 125. On the other hand, Figure 3b shows normalised abundance curves according to hydrolysis times of peptides LQYLYQGPIVL (αS2-CN (111–121)) and YPFPGPIPNSLPQN (β-CN (75–88)), more abundant during kinetics 125. The latter are not or only very weakly present in samples 109-1, -2, -3, considered as the most bitter, suggesting that they do not bring significant bitterness to the samples. They would therefore not be responsible for the difference in bitterness between samples 109-1, -2, -3 and the others.

3.3. Relationship between Generated Peptides and Bitterness during Hydrolysis

3.3.1. Heat Map

The heat map presented Figure 4 shows the differences in terms of normalised abundances between the samples of both kinetics in a more visual way than a table. On the heat map, the peptides are grouped in rows and the samples withdrawn during the hydrolysis kinetics (109 and 125) in columns. The peptides are divided into two groups: A' and B' (left dendrogram) and the samples are divided into two groups: A and B (top dendrogram).

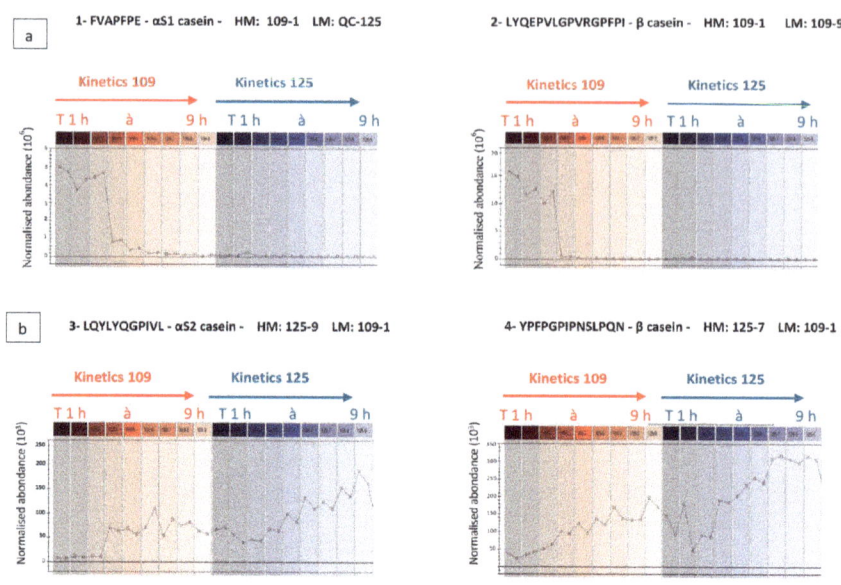

Figure 3. Examples of two opposite evolutions of the normalised abundance of peptides (**a**) whose abundance is highest at the beginning of kinetics 109 and (**b**) whose abundance is highest at the end of kinetics 125. The red and blue colour represent kinetics 109 and kinetics 125, respectively. (HM: highest mean; LM: lowest mean).

In order to identify the peptides responsible for the difference in bitterness, a differential expression analysis was performed by merging kinetics 109 and 125. Among the 18 samples, two different groups were formed according to the bitterness scores obtained by sensory analysis (group #1 and group #2). Group #1 includes the samples 109-1, 109-2, and 109-3, which all had a bitterness intensity greater than 3.91, and group #2 includes the remaining 15 samples of both kinetics with an intensity equal to or lower than 3.91. Among the 116 identified peptides, only 54 are significant ($p \leq 0.05$—noted with an asterisk (*) in Figure 4) which means that they contribute to the difference in terms of bitterness between groups #1 and #2.

The peptide group B' corresponds to the 54 significant peptides discriminated by the differential analysis, and explains the difference in bitterness between the samples in groups A and B. The peptide group A', corresponding to the non-significant peptides of the differential analysis, explains the difference between the samples of kinetics 125 and 109 (without the samples 109-1, 2 and 3). The sample group A corresponds to the three samples 109-1, 109-2 and 109-3 with the highest bitterness intensities according to the sensory data obtained by the panel. These three samples are well differentiated on the heat map: the red colour (corresponding to the peptide group B') and the yellow colour (corresponding to the peptide group A') rectangles on the left show that for the first samples of hydrolysis 109, we have a relatively high normalised abundance of group B' peptides compared to group A'. According to their normalised abundance, these peptides as well as their quantity tend to explain the high bitterness of the first samples at the beginning of the hydrolysis. The second group of samples (B) includes the subgroup of the samples of kinetics 125 in the following order: 125-2, 125-1, 125-3, 125-4, 125-5, 125-8, 125-9, 125-6, and 125-7, and the subgroup of the remaining samples of kinetics 109: 109-4, 109-8, 109-9, 109-5, 109-6 and 109-7. The samples of the kinetics 109 subgroup with the exception of 109-1, 2 and 3 are characterised by high normalised abundances of the A' peptide group. The kinetics 125 subgroup samples are characterised by high normalised abundances of the first eight peptides of group A' and intermediate normalised abundances of the remaining peptides.

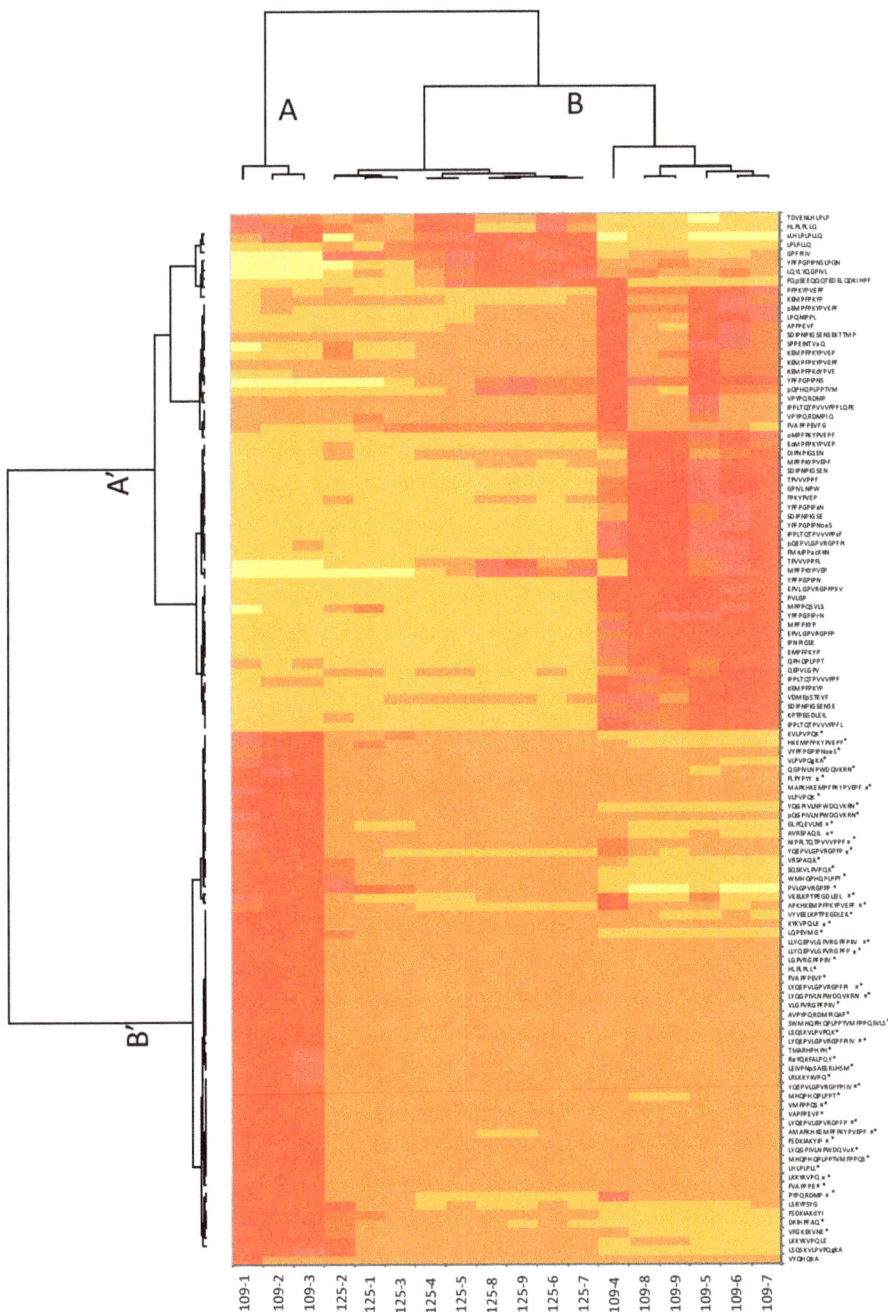

Figure 4. Heat map related to mass spectrometry data. The peptides marked with * correspond to the peptides that explain the significant difference between the two groups of hydrolysates obtained with the differential analysis ($p \leq 0.05$). Peptides marked with an "x" are the peptides identified through random forests as the most influential in explaining the bitterness of hydrolysates.

3.3.2. Regression Trees and Random Forests

RT and RF were also performed by merging the samples of kinetics 109 and 125. The importance of peptides as predictors of the main taste characteristic of enzymatic hydrolysates, namely bitterness, is presented in Figure 5. The measure of this importance quantifies the contribution of each peptide to the prediction of the sensory profile. Based on the confidence interval, we noticed that more than three quarters (94) of the peptides influenced bitterness. In order to build an accurate and parsimonious model and to identify the best subset of predictors from the 94 pre-selected predictor variables, we reduced the number of predictors by adding the peptides sequentially, from the most important to the least important one: all possible subsets from 5 to k variables are considered in turn. For each subset, an RF is constructed with the same parameters as before. The quality of the prediction was computed for each model generated and evaluated using the $RMSE_{OOB}$ index. The lower it is, the better the quality of the prediction will be. The optimal RF was obtained for 22 peptides with a $RMSE_{OOB}$ of 0.3405. The latter therefore corresponds to the first 22 peptides presented in Figure 5.

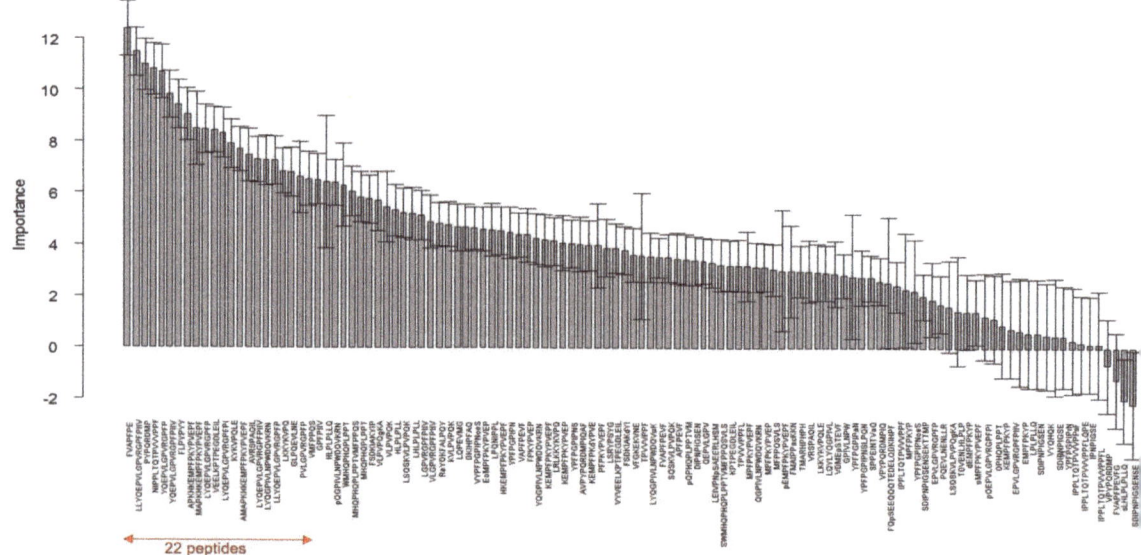

Figure 5. Random forests on the descriptor bitter: importance of the 116 peptides. The confidence intervals (95%) were obtained with 50 random forests of 5000 trees.

The model obtained with these selected 22 peptides was applied to both kinetics for predicting hydrolysate bitterness profiles (Figure 6). Figure 6 represents the evolution of the bitterness intensity of samples during kinetics 109 and 125 over time, with the observed values (full lines associated with full circles and triangles) and the predicted values (dotted lines associated with red empty circles and black triangles). The prediction is quite good with a mean error of 0.34 for the predicted perception of bitterness and a correlation of 0.93 between observed and predicted values.

Thanks to the 22 peptides, an optimal RT has been built (Figure 7). The first peptide which splits the initial 18 samples into 2 groups was FVAPFPE at a normalised abundance threshold value of 2,431,772. The three samples (node 7, $n = 3$) with a value higher than this threshold were considered as the most bitter. These latter are the three first samples of kinetics 109 (109-1, 109-2 and 109-3). On the other hand, fifteen hydrolysates with a FVAPFPE value below this threshold were grouped together (nodes 3, 5 and 6 with, respectively, 6, 8 and 1 sample(s)). Then, when the normalised abundance of NIPPLTQTPVVVPPF was lower than 56,200.67, six samples were separated from the others and revealed the least bitterness perception. A last split was performed another time with the peptide FVAPFPE and beyond the abundance of 656,107.8. The hydrolysate was again considered more bitter when the peptide abundance was higher than this value.

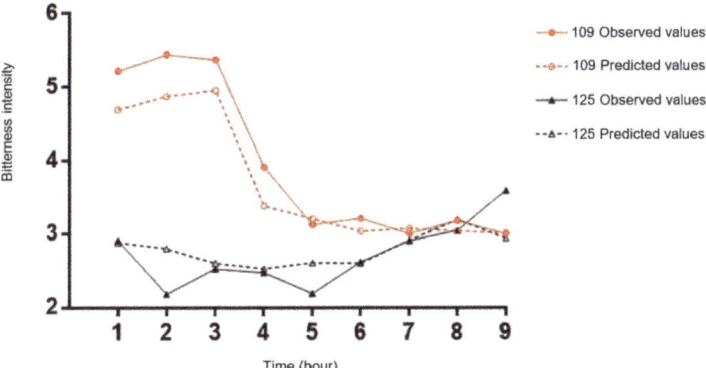

Figure 6. Prediction of the bitterness of out-of-bag (OOB) samples. The red colour represents the evolution of the intensity of samples during kinetics 109 and the black colour represents the samples of kinetics 125. The full lines with full-circles or -triangles represent the observed values and the open-circles and -triangles represent the predicted values.

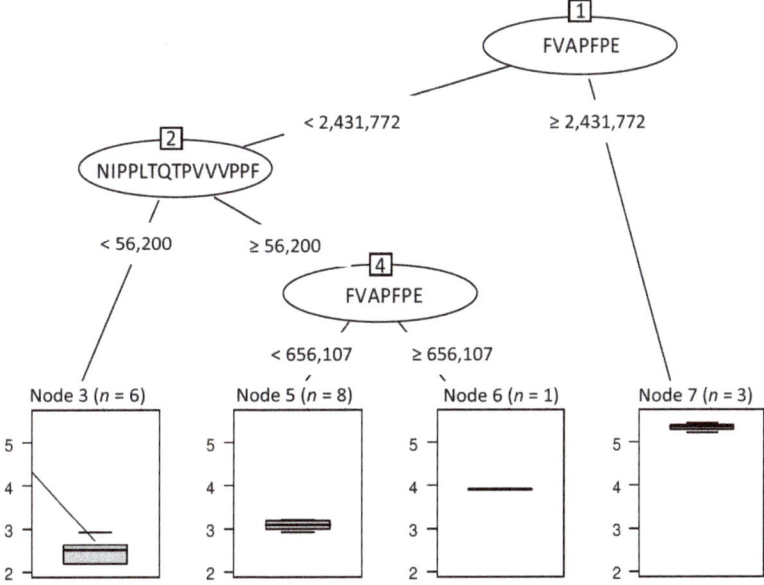

Figure 7. Optimal regression tree built to predict the hydrolysate bitterness from the 22 identified peptides. The boxplots represented below the tree show the intensity and the gradual evolution of bitterness. n = number of hydrolysates for each group defined by a different level of bitterness with sample reference number. Node 3 (n = 6) corresponds to hydrolysates 125-1, -2, -3, -4, -5 and -6; Node 5 (n = 8) corresponds to hydrolysates 125-7, -8, -9/109-5, -6, -7, -8 and -9; Node 6 (n = 1) corresponds to hydrolysate 109-4; Node 7 (n = 3) corresponds to hydrolysates 109-1, -2 and -3.

4. Discussion

The enzymatic hydrolysis of caseins is especially known for the appearance of bitterness, which is a hindrance to their use in the agri-food industry [17]. The first step of this study was to analyse the bitterness profile of two different hydrolysis kinetics called "kinetics 109" and "kinetics 125". For these given hydrolysis conditions, the sensory evaluation, driven with a trained sensory panel, reveals that significantly different bitterness levels are obtained depending on the hydrolysis time. This latter is more visible for kinetics 109, for which we obtained a significant drop in bitterness after about three hours of hydrolysis, reaching an intensity of 3. It is well known that the development of a specific sensory profile in protein hydrolysates depends on the protein source, enzyme specificity and the conditions of the hydrolysis [18,19]. Based on all these rules, we selected and combined some specific enzymes known in the literature to aid in the development of low bitterness hydrolysates for applications in nutrition and the agri-food market. These enzymes are Flavorpro 937MDP ™, a mixture of endo- and exoproteases, which according to the manufacturer Biocatalysts, has been developed to remove the excessive bitter-tasting peptides produced when using animal and bacterial proteases; Promod 523MDP ™, an endoprotease with a bromelain activity, which is effective in the production of highly digestible proteins [20]; Protamex, which also substantially reduces bitterness when hydrolysing caseins, as indicated by the company Novozymes [21]; and finally Flavourzyme, which contains both endo- and exoprotease activities and has shown its efficiency in obtaining less bitter milk protein hydrolysates compared to other enzyme preparations [22]. This efficiency has been linked to the presence of high exopeptidase activity within this preparation [23,24]. Indeed, the exopeptidases cleave peptides from their C- or N-terminal extremities, allowing a reduction in the bitterness due to terminal hydrophobic amino acid residues, for example [25].

The sensory study of kinetics 109 and 125 allowed us to monitor on one hand the evolution of certain sensory features, and on the other hand the physico-chemical parameters, such as the DH. Concerning the latter, it emerged that it was not always correlated with the bitterness of the hydrolysates, as already confirmed by a lot of studies [26–29].

The second step was to identify the peptides generated during the time course of hydrolyses and concomitantly quantify their normalised abundance. These mass spectrometry data were then analysed by performing a heat map combined with a differential expression analysis of both kinetics combined. These statistical tests are very often used in the presence of big data such as OMICS-type data [30]. This heat map brings an overview of the peptide abundance evolution during hydrolysis kinetics and an image of the proximity of the samples. For kinetics 109, the lower peptide cluster is very abundant at the beginning of hydrolysis, then decreases over time while remaining stable during the last hours of the hydrolysis. Combined with the differential expression, 54 peptides have been identified as responsible for the bitterness difference. These latter are the most abundant peptides in the first three bitterest samples 109-1, 109-2 and 109-3, except LKKYKVPQLE, VYQHQKA and LSQSKVLPVPQgKA. The evolution of peptide abundance during kinetics 125 seems to be almost identical for all the samples constituting it, except for samples 125-1, -2, -3, which would explain the stability of its bitterness over time.

One of the main strengths of this study concerns the use of RT and RF methodologies to predict the bitterness of samples. It is important to note that single trees are easy to interpret and to establish the relationships inside the dataset. However, they are unstable, and small perturbations in the dataset can completely change their structure. Therefore, for the prediction of the sample taste, the use of the whole RF is more convenient. In fact, we obtained a very satisfactory quality index with a value of 0.34, meaning that the bitterness intensity can be predicted with a confidence interval of 0.34. Such a value is a good estimation in sensory evaluation. Moreover, the results showed that the RF highlighted the importance of peptides in explaining the bitterness of the 18 samples. As shown in Figure 5, the 22 most influential peptides selected to construct the RF are among the 54 peptides most involved in the differentiation of the two sample categories #1 and #2 derived from differential expression. The simultaneous presence of this set of peptides and their abundance are the cause of the difference in bitterness existing between the samples. An additional verification was made with the BIOPEP database. Available online: http://www.uwm.edu.pl/biochemia/index.php/en/biopep (accessed on 10 February 2021) [31] and the literature to determine if some of those 22 peptides were already reported as being bitter. In this database, all data about the taste of the peptides were obtained from sensory studies described in the literature. The peptides YQEPVLGPVRGPFP, YQEPVL-GPVRGPFPIIV, APKHKEMPFPKYPVEPF, MAPKHKEMPFPKYPVEPF, AMAPKHKEMPF-PKYPVEP and PVLGPVRGPFP had already been identified [17,32–34]. The studies of Karametsi et al. highlighted the following bitter peptides: GPVRGPFPIIV and YQEPVL-GPVRGPFPI [33], and those of Toelstede et al. the sequence GPVRGPFP [35]. These results show that their primary structure is similar to peptides LLYQEPVLGPVRGPFPIIV, LYQEPVLGPVRGPFPI, LYQEPVLGPVRGPFP, LYQEPVLGPVRGPFPIIV and LLYQEPVL-GPVRGPFP that we have identified in this study, suggesting that this protein region of the β-casein is conducive to the release of bitter peptides. Besides that, the FVAPFPE peptide is already known to display angiotensin converting enzyme inhibitory activity [36], but no information on its taste has been given. However, a peptide of close structure "FFVAPF-PEVFGK", referenced in the BIOPEP database, has been identified as bitter. Moreover, a correlation between the bitter taste of a peptide and bioactivity has been demonstrated [19]. The VEELKPTPEGDLEIL peptide was also characterised as bitter by Spellman et al. (2015). No information was found on the other peptide sequences. Thus, a large majority of peptides highlighted by RFs have already been identified in the literature. However, it is worth to note that it is the combination of the presence and/or the absence and the association of different peptides which is responsible for hydrolysate's bitterness.

However, this prediction model has a major limitation since it can only be used with the same enzymes used in this study. The use of RFs makes it possible to take into account the entire complex mixture represented by a peptide hydrolysate. Indeed, the synergistic and antagonistic effects that may exist between peptides and their impact on bitterness are considered.

5. Conclusions

The impact of bitterness on food rejection has been studied extensively. Therefore, the development of protein hydrolysates with low levels of bitterness is an essential challenge for their incorporation in various foods. The hydrolysates formulated here may be used in the development of future food formulations such as peptide-fortified ready-to-drink infant formulae and low pH beverages such as fruit juices, for instance.

The data generated may be employed to inform the selection of a certain type of enzyme preparation and target the degree of hydrolysis values to generate hydrolysates with adequate sensory properties. A peptide hydrolysate is a complex mixture where interactions between peptides can complicate their study. However, random forests appear to be a useful tool for their analysis. Moreover, the use of RT and RF methodologies allows us, on one side, to highlight peptides involved in the explanation of the bitterness of samples, and on the other side to predict the bitterness profile of a micellar casein hydrolysate.

Supplementary Materials: The following are available online at https://www.mdpi.com/article/10.3390/foods10061312/s1, Table S1: list of peptides identified by the HPLC-MS/MS analyses.

Author Contributions: D.D.: investigation, formal analysis, visualization, data curation, writing—original draft; B.D.: investigation, formal analysis, validation, data curation, supervision, writing—original draft; P.C.: investigation, formal analysis, validation, data curation, writing—original draft; A.B.: methodology, supervision, writing—reviewing and editing; S.C.: methodology, formal analysis, data curation, supervision, funding acquisition, writing—reviewing and editing; R.F.: methodology, supervision, funding acquisition, writing—reviewing and editing; C.F.: conceptualization, methodology, visualization, validation, data curation, supervision, funding acquisition, writing—reviewing and editing. All authors have read and agreed to the published version of the manuscript.

Funding: This work has been carried out in the framework of the joint laboratory between the Charles Viollette institute and the Ingredia project (Allinpep) and in the framework of the Alibiotech research program which is financed by the European Union, French State and the French Region of Hauts-de-France. The HPLC-MS/MS experiments were performed on the REALCAT platform funded by a French governmental subsidy managed by the French National Research Agency (ANR) within the frame of the "Future Investments'program (ANR-11- EQPX-0037)". The Hauts-de-France region and the FEDER, the Ecole Centrale de Lille and the Centrale Initiatives Foundation are also warmly acknowledged for their financial contributions to the acquisition of REALCAT platform equipment.

Institutional Review Board Statement: Not applicable.

Informed Consent Statement: Informed consent was obtained from all subjects involved in the study.

Data Availability Statement: Data is contained within the article or supplementary material.

Conflicts of Interest: The project was funded by Ingredia S.A. Dahlia Daher and Alain Baniel are employed by Ingredia S.A.

References

1. Manninen, A.H. Protein hydrolysates in sports nutrition. *Nutr. Metab.* **2009**, *6*, 38. [CrossRef] [PubMed]
2. Fujita, S.; Glynn, E.L.; Timmerman, K.L.; Rasmussen, B.B.; Volpi, E. Supraphysiological hyperinsulinaemia is necessary to stimulate skeletal muscle protein anabolism in older adults: Evidence of a true age-related insulin resistance of muscle protein metabolism. *Diabetologia* **2009**, *52*, 1889–1898. [CrossRef] [PubMed]
3. Murray, T.K.; Baker, B.E. Studies on protein hydrolysis. I.—Preliminary observations on the taste of enzymic protein-hydrolysates. *J. Sci. Food Agric.* **1952**, *3*, 470–475. [CrossRef]
4. Ichikawa, K.; Yamamoto, T.; Fukomoto, J. Bitter-tasting Peptides Produced by Proteinases. I. The Formation of Bitter-tasting Peptides by the Neutral Proteinase of Bacillus Subtilis and Isolation of the Peptides. *J. Agric. Chem. Soc.* **1959**, *33*, 1044.
5. Sebald, K.; Dunkel, A.; Schäfer, J.; Hinrichs, J.; Hofmann, T. Sensoproteomics: A New Approach for the Identification of Taste-Active Peptides in Fermented Foods. *J. Agric. Food Chem.* **2018**, *66*, 11092–11104. [CrossRef]
6. Fujimaki, M.; Yamashita, M.; Arai, S.; Kato, H.; Gonda, M. Enzymatic Modification of Proteins in Foodstuffs: Part I. Enzymatic Proteolysis and Plastein Synthesis Application for Preparing Bland Protein-like Substances Part II. Nutritive Properties of Soy Plastein and its Bio-utility Evaluation in Rats. *Agric. Biol. Chem.* **1970**, *34*, 1325–1337. [CrossRef]
7. Matoba, T.; Hayashi, R.; Hata, T. Isolation of Bitter Peptides from Tryptic Hydrolysate of Casein and their Chemical Structure. *Agric. Biol. Chem.* **1970**, *34*, 1235–1243. [CrossRef]
8. Kim, I.; Kawamura, Y.; Lee, C.-H. Isolation and Identification of Bitter Peptides of Tryptic Hydrolysate of Soybean 11S Glycinin by Reverse-phase High-performance Liquid Chromatography. *J. Food Sci.* **2006**, *68*, 2416–2422. [CrossRef]
9. Iwaniak, A.; Hrynkiewicz, M.; Bucholska, J.; Minkiewicz, P. Understanding the nature of bitter-taste di- and tripeptides derived from food proteins based on chemometric analysis. *J. Food Biochem.* **2018**, 1–7. [CrossRef]
10. Daher, D.; Deracinois, B.; Baniel, A.; Wattez, E.; Dantin, J.; Froidevaux, R.; Chollet, S.; Flahaut, C. Principal Component Analysis from Mass Spectrometry Data Combined to a Sensory Evaluation as a Suitable Method for Assessing Bitterness of Enzymatic Hydrolysates Produced from Micellar Casein Proteins. *Foods* **2020**, *9*, 1354. [CrossRef]
11. Nielsen, P.M.; Petersen, D.; Dambmann, C. Improved Method for Determining Food Protein Degree of Hydrolysis. *J. Food Sci.* **2001**, *66*, 642–646. [CrossRef]
12. Guo, H.; Zhang, W.; Ni, C.; Cai, Z.; Chen, S.; Huang, X. Heat map visualization for electrocardiogram data analysis. *BMC Cardiovasc. Disord.* **2020**, *20*, 277. [CrossRef] [PubMed]
13. Breiman, L. Random Forests. *Mach. Learn.* **2001**, *45*, 5–32. [CrossRef]
14. R Core Team. *R: A Language and Environment for Statistical Computing*; R Foundation for Statistical Computing; R Core Team: Vienne, France, 2018.
15. Therneau, R.; Atkinson, B.; Ripley, B. *Rpart: Recursive Partitioning and Regression Trees*; 2019. Available online: https://cran.r-project.org/web/packages/rpart/rpart.pdf (accessed on 7 July 2020).
16. Liaw, A.; Wiener, M. Package "randomForest": Breiman and Cutler's random forests for classification and regression. *R Dev. Core Team* **2018**, *4*, 6–10.
17. Lemieux, L.; Simard, R.E. Bitter flavour in dairy products. II. A review of bitter peptides from caseins: Their formation, isolation and identification, structure masking and inhibition. *Le Lait* **1992**, *72*, 335–385. [CrossRef]
18. Tanimoto, S.; Watanabe, M.; Arai, S. Bitter Flavor of Protein Hydrolysates and Synthetic Peptides. In *Developments in Food Science*; Charalambous, G., Ed.; Off-Flavors in Foods and Beverages; Elsevier: Amsterdam, The Netherlands, 1992; Volume 28, pp. 547–566.
19. Murray, N.M.; O'Riordan, D.; Jacquier, J.-C.; O'Sullivan, M.; Holton, T.A.; Wynne, K.; Robinson, R.C.; Barile, D.; Nielsen, S.D.; Dallas, D.C. Peptidomic screening of bitter and nonbitter casein hydrolysate fractions for insulinogenic peptides. *J. Dairy Sci.* **2018**, *101*, 2826–2837. [CrossRef] [PubMed]

30. Pavan, R.; Jain, S.; Shraddha; Kumar, A. Properties and therapeutic application of bromelain: A review. *Biotechnol. Res. Int.* **2012**, *2012*, 976203. [CrossRef] [PubMed]
31. Cheung, I.W.Y.; Li-Chan, E.C.Y. Application of taste sensing system for characterisation of enzymatic hydrolysates from shrimp processing by-products. *Food Chem.* **2014**, *145*, 1076–1085. [CrossRef]
32. Uluko, H.; Zhang, S.; Liu, L.; Chen, J.; Sun, Y.; Su, Y.; Li, H.; Cui, W.; Lv, J. Effects of microwave and ultrasound pretreatments on enzymolysis of milk protein concentrate with different enzymes. *Int. J. Food Sci. Technol.* **2013**, *48*, 2250–2257. [CrossRef]
33. Saha, B.C.; Hayashi, K. Debittering of protein hydrolyzates. *Biotechnol. Adv.* **2001**, *19*, 355–370. [CrossRef]
34. FitzGerald, R.; O'Cuinn, G. Enzymatic debittering of food protein hydrolysates. *Biotechnol. Adv.* **2006**, *24*, 234–237. [CrossRef] [PubMed]
35. Cheung, L.K.Y.; Aluko, R.E.; Cliff, M.A.; Li-Chan, E.C.Y. Effects of exopeptidase treatment on antihypertensive activity and taste attributes of enzymatic whey protein hydrolysates. *J. Funct. Foods* **2015**, *13*, 262–275. [CrossRef]
36. Slattery, H.; Fitzgerald, R.J. Functional Properties and Bitterness of Sodium Caseinate Hydrolysates Prepared with a Bacillus Proteinase. *J. Food Sci.* **1998**, *63*, 418–422. [CrossRef]
37. Barry, C.M.; O'Cuinn, G.; Harrington, D.; O'Callaghan, D.M.; Fitzgerald, R.J. Debittering of a Tryptic Digest of Bovine p-casein Using Porcine Kidney General Aminopeptidase and X-Prolydipeptidyl Aminopeptidase from Lactococcus lactis subsp. cremoris AM2. *J. Food Sci.* **2000**, *65*, 1145–1150. [CrossRef]
38. Bouchier, P.J.; O'Cuinn, G.; Harrington, D.; FitzGerald, R.J. Debittering and Hydrolysis of a Tryptic Hydrolysate of β-casein with Purified General and Proline Specific Aminopeptidases from Lactococcus lactis ssp. cremoris AM2. *J. Food Sci.* **2001**, *66*, 816–820. [CrossRef]
39. O'Sullivan, D.; Nongonierma, A.B.; FitzGerald, R.J. Bitterness in sodium caseinate hydrolysates: Role of enzyme preparation and degree of hydrolysis. *J. Sci. Food Agric.* **2017**, *97*, 4652–4655. [CrossRef]
40. Williams, J.R.; Yang, R.; Clifford, J.L.; Watson, D.; Campbell, R.; Getnet, D.; Kumar, R.; Hammamieh, R.; Jett, M. Functional Heatmap: An automated and interactive pattern recognition tool to integrate time with multi-omics assays. *BMC Bioinform.* **2019**, *20*, 81. [CrossRef]
41. Minkiewicz, P.; Iwaniak, A.; Darewicz, M. BIOPEP-UWM Database of Bioactive Peptides: Current Opportunities. *Int. J. Mol. Sci.* **2019**, *20*, 5978. [CrossRef]
42. Tamura, M.; Miyoshi, T.; Mori, N.; Kinomura, K.; Kawaguchi, M.; Ishibashi, N.; Okai, H. Mechanism for the Bitter Tasting Potency of Peptides Using O-Aminoacyl Sugars as Model Compounds. *Agric. Biol. Chem.* **1990**, *54*, 1401–1409. [CrossRef]
43. Karametsi, K.; Kokkinidou, S.; Ronningen, I.; Peterson, D.G. Identification of Bitter Peptides in Aged Cheddar Cheese. *J. Agric. Food Chem.* **2014**, *62*, 8034–8041. [CrossRef]
44. Singh, T.K.; Young, N.D.; Drake, M.; Cadwallader, K.R. Production and Sensory Characterization of a Bitter Peptide from β-Casein. *J. Agric. Food Chem.* **2005**, *53*, 1185–1189. [CrossRef] [PubMed]
45. Toelstede, S.; Hofmann, T. Sensomics Mapping and Identification of the Key Bitter Metabolites in Gouda Cheese. *J. Agric. Food Chem.* **2008**, *56*, 2795–2804. [CrossRef] [PubMed]
46. Durak, M.Z.; Turan, N.A. Antihypertensive Peptides in Dairy Products. Available online: https://biomedgrid.com/index.php (accessed on 25 January 2021).

Article

Proteomic Characterization of Bacteriophage Peptides from the Mastitis Producer *Staphylococcus aureus* by LC-ESI-MS/MS and the Bacteriophage Phylogenomic Analysis

Ana G. Abril [1], Mónica Carrera [2,*], Karola Böhme [3], Jorge Barros-Velázquez [4], Benito Cañas [5], José-Luis R. Rama [1], Tomás G. Villa [1] and Pilar Calo-Mata [4,*]

1. Department of Microbiology and Parasitology, Faculty of Pharmacy, University of Santiago de Compostela, 15898 Santiago de Compostela, Spain; anagonzalezabril@hotmail.com (A.G.A.); joserodrama@gmail.com (J.-L.R.R.); tomas.gonzalez@usc.es (T.G.V.)
2. Department of Food Technology, Spanish National Research Council, Marine Research Institute, 36208 Vigo, Spain
3. Agroalimentary Technological Center of Lugo, 27002 Lugo, Spain; KarolaBoehme@gmx.de
4. Department of Analytical Chemistry, Nutrition and Food Science, School of Veterinary Sciences, University of Santiago de Compostela, 27002 Lugo, Spain; jorge.barros@usc.es
5. Department of Analytical Chemistry, Complutense University of Madrid, 28040 Madrid, Spain; bcanas@quim.ucm.es
* Correspondence: mcarrera@iim.csic.es (M.C.); p.calo.mata@usc.es (P.C.-M.)

Abstract: The present work describes LC-ESI-MS/MS MS (liquid chromatography-electrospray ionization-tandem mass spectrometry) analyses of tryptic digestion peptides from phages that infect mastitis-causing *Staphylococcus aureus* isolated from dairy products. A total of 1933 nonredundant peptides belonging to 1282 proteins were identified and analyzed. Among them, 79 staphylococcal peptides from phages were confirmed. These peptides belong to proteins such as phage repressors, structural phage proteins, uncharacterized phage proteins and complement inhibitors. Moreover, eighteen of the phage origin peptides found were specific to *S. aureus* strains. These diagnostic peptides could be useful for the identification and characterization of *S. aureus* strains that cause mastitis. Furthermore, a study of bacteriophage phylogeny and the relationship among the identified phage peptides and the bacteria they infect was also performed. The results show the specific peptides that are present in closely related phages and the existing links between bacteriophage phylogeny and the respective *Staphylococcus* spp. infected.

Keywords: pathogen detection; LC-ESI-MS/MS; proteomics; mass spectrometry; phage peptide biomarker

1. Introduction

The vast majority of mastitis cases are due to an intramammary infection caused by a microorganism belonging to either the *Staphylococcus* or *Streptococcus* genus [1,2]. *Staphylococcus aureus* is considered one of the major foodborne pathogens that can cause serious food intoxication in humans due to the production of endotoxins; this pathogen remains a major issue in the dairy industry due to its persistence in cows, its pathogenicity, its contagiousness and its ease of colonization of the skin and mucosal epithelia [3–5].

It is well-known that *S. aureus* bacteriophages encode genes for staphylococcal virulence factors, such as Panton-Valentine leucocidin, staphylokinase, enterotoxins, chemotaxis-inhibitory proteins or exfoliative toxins [6]. These phages are usually integrated into bacterial chromosomes as prophages, wherein they encode new properties in the host, or vice versa, as transcriptions may hardly be affected by gene disruptions [7]. Phage-encoded recombinases, rather than the host recombinase, RecA, are involved in bacterial genome excisions and integrations [8,9]. These integrations may occur at specific bacterial genome sites that are identical to those present in the DNA of the phage, or, as in the case of phage

Mu (as long as the given gene is not expressed), some phages can integrate randomly within the bacterial genome. In addition, bacteriophage and staphylococcal species interactions may substantially alter the variability of the bacterial population [10,11].

All known *S. aureus* phages are composed of an icosahedral capsid filled with double stranded DNA and a thin, filamentous tail, and they belong to the order *Caudovirales* (tailed phages) [12,13]. Some *Podoviridae* family phages, such as the *Staphylococcus* viruses S13 and S24-1, have been reported, characterized and used in phage therapy against *S. aureus* infections [14]. There are some well-known *Siphoviridae* phages of *S. aureus*, such as the prophage φSaBov, which is integrated into a bovine mastitis-causing *S. aureus* strain [15].

The interaction between bacteria and bacteriophages leads to an exchange of genetic information, which enables bacteria to rapidly adapt to challenging environmental conditions and to be highly dynamic [11,16]. As closely related phages normally occupy the same genome location in different bacteria, a specific site in different bacterial strains can be occupied by completely different phages or can be empty.

Conventional culture-based methods have been used for the detection of pathogenic bacteria [17,18] and their phages [19,20]; however, at this point, these procedures are time consuming and laborious. For this reason, new, rapid molecular microbial diagnostic methods based on genomics and proteomics tools have been developed to achieve faster and more efficient bacterial and bacteriophage identification [1,21–24]. Specifically, phage typing is a classic technique for such purposes [25]. Moreover, biosensors based on phage nucleic acids, receptor-binding proteins (RBPs), antibodies and phage display peptides (PDPs) have been used for pathogen detection [26–30].

Mass spectrometry techniques, such as MALDI-TOF MS (matrix-assisted laser desorption/ionization time-of-flight mass spectrometry) and LC-ESI-MS/MS (liquid chromatography electrospray ionization-tandem mass spectrometry), have been used for the analysis and detection of specific diagnostic peptides in pathogenic bacterial strains [31,32]. In addition, LC-ESI-MS/MS methods have been employed for the identification and detection of bacteriophages [19]. In the case of bacteriophage detection and identification by a mass spectrometry analysis, the required production of viruses may be time-consuming. The detection of prophages based on protein biomarkers can be an alternative to genomic detection, and in this sense, proteomic techniques can be cheaper and faster and can ascertain different bacteriophage species by using a single analysis [33]. Based on the specificity of many bacteriophages with their hosts, bacteriophages are considered signal amplifiers; therefore, the detection of peptides from phages is suitable for pathogen identification. For example, Serafim et al. 2017 [33] identified bacteriophage lambda by a LC-ESI-MS/MS analysis. Moreover, the identification of peptides by means of LC-ESI-MS/MS from bacteriophage-infected *Streptococcus* has been performed, which revealed new information on phage phylogenomics and their interactions with the bacteria they infect [19]. However, no study has been published on *S. aureus* phage detection and identification by LC-ESI-MS/MS or on *S. aureus* phage characterization without a previous phage purification step. Viral genomic detection and phage display are time-consuming methods. Here, we describe an easy, fast and accurate method for the detection of bacteriophages without the need for the pretreatment of bacterial lysis for bacteriophage replication. This method led to the identification of putative temperate and virulent phages present in the analyzed strains.

A previously published work performed by our laboratory [3] studied the global proteome of several strains of *S. aureus* by shotgun proteomics. Important virulence protein factors and functional pathways were characterized by a protein network analysis. In this work, and for the first time, we aimed to use proteomics to characterize phage contents in different *S. aureus* strains to identify the relevant phage-specific peptides of several *S. aureus* strains and to identify both phages and bacterial strains by LC-ESI-MS/MS.

2. Materials and Methods

2.1. Bacteria

In this study, a total of 20 different *S. aureus* strains obtained from different sources were analyzed (Table S1 in Supplemental Data 2). These strains were previously characterized by MALDI-TOF mass spectrometry [1] after being obtained from the Institute of Science of Food Production of the National Research Council of Italy (Italy) and from the Spanish Type Culture Collection (Spain). The majority of the strains are from food origins, except for strain U17, which is a human clinical strain. Strains ATCC (American Type Culture Collection) 9144 and ATCC 29213 are classified as *S. aureus* subsp. *aureus*, while strain ATCC 35845 is categorized as *S. aureus* subsp. *anaerobius*. In previous works, the species identification of *S. aureus* and the presence of enterotoxins were evaluated by multiplex polymerase chain reactions (multiplex PCRs) [3,34,35]. The strains were reactivated in a brain–heart infusion medium (BHI, Oxoid Ltd., Hampshire, UK) and incubated at 31 °C for 24 h. Bacterial cultures were then grown on plate count agar (PCA, Oxoid) at 31 °C for 24 h [1,3,36]. Tubes of broth were inoculated under aerobic conditions.

2.2. Protein Extraction and Peptide Sample Preparation

Protein extraction was prepared as described previously [37]. All analyses were performed in triplicate. Protein extracts were subjected to in-solution tryptic digestion [38].

2.3. Shotgun LC-MS/MS Analysis

Peptide digests were acidified with formic acid (FA), cleaned on a C18 MicroSpin™ column (The Nest Group, South-borough, MA, USA) and analyzed by LC-ESI-MS/MS using a Proxeon EASY-nLC II Nanoflow system (Thermo Fisher Scientific, San Jose, CA, USA) coupled to an LTQ-Orbitrap XL mass spectrometer (Thermo Fisher Scientific, San Jose, CA, USA) [3]. Peptide separation (2 µg) was performed on a reverse-phase (RP) column (EASY-Spray column, 50 cm × 75 µm ID, PepMap C18, 2-µm particles, 100-Å pore size, Thermo Fisher Scientific, San Jose, CA, USA) with a 10-mm precolumn (Accucore XL C18, Thermo Fisher Scientific, San Jose, CA, USA) using a linear 120-min gradient from 5% to 35% solvent B (solvent A: 98% water, 2% ACN (Acetonitrile) and 0.1% FA and solvent B: 98% ACN, 2% water and 0.1% FA) at a flow rate of 300 nL/min. For ionization, a spray voltage of 1.95 kV and a capillary temperature of 230 °C were used. Peptides were analyzed in the positive mode from 400 to 1600 amu (1 µscan), which was followed by 10 data-dependent collision-induced dissociation (CID) MS/MS scans (1 µscan) using an isolation width of 3 amu and a normalized collision energy of 35%. Fragmented masses were set in dynamic exclusion for 30 s after the second fragmentation event, and unassigned charged ions were excluded from the MS/MS analysis.

2.4. LC-MS/MS Mass Spectrometry Data Processing

LC-ESI-MS/MS spectra were searched using SEQUEST-HT (Proteome Discoverer 2.4, Thermo Fisher Scientific, San Jose, CA, USA) against the *S. aureus* UniProt/TrEMBL database (208,158 protein sequence entries in July 2020). The following parameters were used: semi-tryptic cleavage with up to two missed cleavage sites and tolerance windows set at 10 ppm for the precursor ions and 0.06 Da for the MS/MS fragment ions. These additional identified semi-tryptic peptides increased the sequence coverage and confidence in protein assignments. The variable modifications that were allowed were as follows: (M*) methionine oxidation (+15.99 Da), (C*) carbamidomethylation of Cys (+57.02 Da) and acetylation of the N-terminus of the protein (+42.0106 Da). To validate the peptide assignments, the results were subjected to a statistical analysis with the Percolator algorithm [39]. The false discovery rate (FDR) was kept below 1%. The mass spectrometric data were deposited into the public database PRIDE (Proteomics Identification Database), with the dataset identifier PXD023530.

2.5. Selection of Potential Peptide Biomarkers

For each peptide identified by LC-ESI-MS/MS, we used the BLASTp program to determine the homologies and exclusiveness with protein sequences registered in the NCBI (National Center for Biotechnology Information) database [40]. For the BLASTp search, the *Staphylococcus* taxon was included and excluded with the aim of finding the peptides that belonged to the *Staphylococcus* phages, *Staphylococcus* spp. and only to *S. aureus*.

2.6. Phage Genome Comparison and Relatedness

Genomes of all studied *Staphylococcus* spp. phages were downloaded from the GenBank database, analyzed and compared using the Web server VICTOR (Virus Classification and Tree Building Online Resource, http://ggdc.dsmz.de/victor.php, accessed on 27 November 2020) for the calculation of the intergenomic distances and the construction of the phylogenomic tree [41].

3. Results

3.1. S. aureus Proteome Repository

Protein mixtures from each of the 20 different *S. aureus* strains (Table S1 in Supplemental Data 2) were digested with trypsin and analyzed by LC-ESI-MS/MS.

A total of 1933 nonredundant peptides corresponding to 1282 nonredundant annotated proteins were identified for all *S. aureus* strains (see the Excel dataset in Supplemental Data 1). Among them, 79 phage peptides were identified. These peptides belong to proteins such as phage repressors, structural phage proteins, uncharacterized phage proteins and complement inhibitors. Figure 1 shows a comparative representation of the different types of phage proteins identified in this study. These phage peptides were selected and analyzed using the BLASTp algorithm. For the BLASTp search, *Staphylococcus* was included and excluded with the aim of finding peptides belonging to *Staphylococcus* bacteriophages.

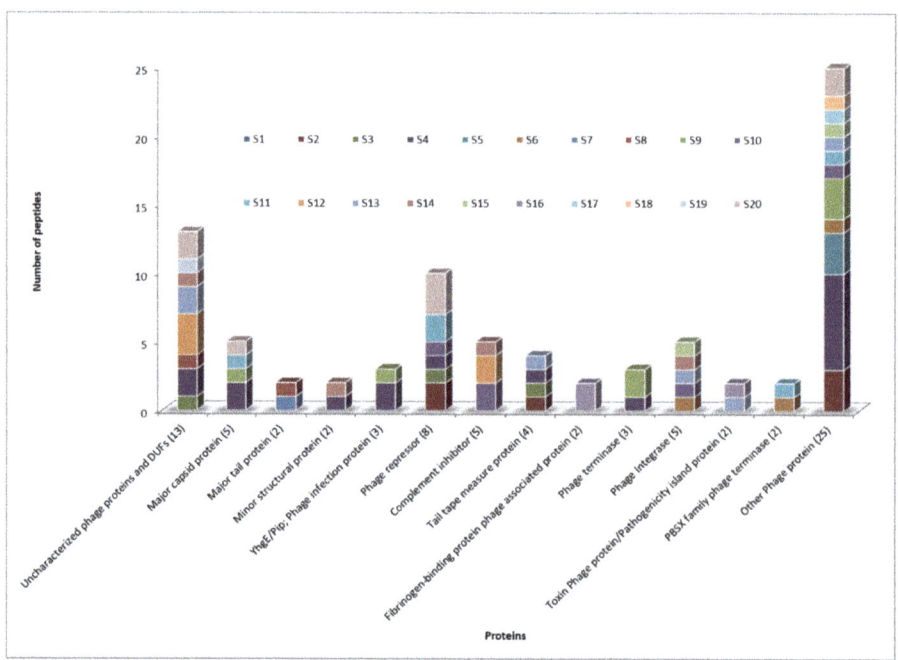

Figure 1. Comparative representation of different types of phage proteins identified in this study for the different strains (represented by different colors). The number of each type of protein is shown in parentheses.

The obtained staphylococcal phage-specific peptides shared homology with the *Staphylococcus* phages and *Staphylococcus* spp. in the NCBI database. Among them, all shared homology with *S. aureus*; however, eighteen peptides were specific to *S. aureus* (IRLPYYDVK, LYVGVFNPEATK, SIINGKLDSQWTVPNEHK, M*NDSNQGLQANPQYTIHYLSQEITR, PCPALM*NKRNSIATIHR, SQDSNLTPELSTKAPK, ESINANTYINQNLEK, VAVLSTPLVTS-FESK, KDGEILFDAIDIYLRNK, MPVYKDGNTGKWYFSI, KTTSEALKEVLSDT, EPKPV-DATGADDPLKPDDRM*ITNFHANLVDQKVSY, MSHNALTTGIGIGAGAG, VQHPGK-LVNKVM*SGLNINFGGGANATAK, QM*MEGLSGVMDLAAVSGEDLGAVSDIVTDGLTA FGLKAKDSG, KSNVEAFSNAVK, GMVASMQMQVVQVNVLTM*ELAQQNAMLTQQLTELK and DIITVYC*PENGTATDEY). Figure S1 shows the MS/MS spectra for these *S. aureus*-specific peptide biomarkers. Table 1 summarizes the list of 79 specific staphylococcal bacteriophage peptides, bacterial peptides with putative phage origins and bacteria and phages with 100% homology with respect to the NCBI protein database.

All staphylococcal phage peptides with 100% homology were found to belong to the *Siphoviridae* family: 52 staphylococcal phages belong to the *Phietavirus* genus, 37 belong to the *Biseptimavirus* genus, 30 are *Triavirus*, two are phieta-like viruses and one is a SPbeta-like virus, and the others are nonclassified *Siphoviridae* viruses (Table S2 in Supplemental Data 2). *Siphoviridae* genomes are usually organized into functional modules, such as lysogeny, DNA replication, packaging, morphogenesis and lysis modules [6,42].

Table 1. Phage origin peptides identified in *Staphylococcus aureus* strains. NCBI (National Center for Biotechnology Information).

Strain	Protein	Peptide	Bacteria with 100% Homology Based on the NCBI Protein Database	Phages with 100% Homology Based on the NCBI Protein Database
S4	Uncharacterized phage protein	IRLPYYDVK	*Staphylococcus aureus*	*Staphylococcus* phage StauST398-2
S4	Uncharacterized phage protein	AVAELLKEINR	*Staphylococcus argenteus* *Staphylococcus simiae* *Staphylococcus aureus*	*Staphylococcus* virus 71 *Staphylococcus* virus 55 *Staphylococcus* virus 88
S4	Major capsid protein	LLHALPTGNDSGGDKLLPK	*Staphylococcus aureus* *Staphylococcus xylosus* *Staphylococcus muscae* *Staphylococcus haemolyticus* *Staphylococcus argenteus* *Streptococcus pneumoniae*	*Staphylococcus* phage phiSa2wa_st72 *Staphylococcus* phage phiSa2wa_st121mssa *Staphylococcus* phage vB_SauS_phi2 *Staphylococcus* phage StauST398-2 *Staphylococcus* phage LH1 *Staphylococcus* phage phiSa2wa_st30 *Staphylococcus* virus phi12 *Staphylococcus* virus 3a *Staphylococcus* virus phiSLT *Staphylococcus* phage tp310-2 *Staphylococcus* phage vB_SauS_JS02 *Staphylococcus* phage R4 *Staphylococcus* phage vB_SauS_fPfSau02 *Staphylococcus* phage SA137ruMSSAST121PVL
S4	Major capsid protein	RVSYTLDDDDFITDVETAKELKL	*Staphylococcus aureus* 12S01399 *Staphylococcus aureus* *Staphylococcus aureus* A9299 *Staphylococcus aureus* A9765 *Staphylococcus argenteus* *Staphylococcus aureus* A6300 *Staphylococcus* sp. Terrabacteria group *Escherichia coli*	*Staphylococcus* phage LH1 *Staphylococcus* phage StauST398-2 *Staphylococcus* phage vB_SauS_phi2 *Staphylococcus* phage R4
S7	Major tail protein	LYVGVFNPEATK	*Staphylococcus aureus*	*Staphylococcus* phage vB_SauS_ phi2 *Staphylococcus* virus phi12 *Staphylococcus* virus phiSLT *Staphylococcus* phage R4 *Staphylococcus* phage vB_SauS_JS02 *Staphylococcus* phage SH-St 15644 *Staphylococcus* virus 3a *Staphylococcus* phage P240
S8	Uncharacterized phage protein	M*NDSNQGLQANPQYTIHYLSQEITR	*Staphylococcus aureus*	*Staphylococcus* phage phiN315

Table 1. Cont.

Strain	Protein	Peptide	Bacteria with 100% Homology Based on the NCBI Protein Database	Phages with 100% Homology Based on the NCBI Protein Database
S8	Major tail protein	AYINITGLGFAK	Staphylococcus aureus Staphylococcus argenteus Pararheinheimera mesophila	Staphylococcus phage phiNM3 Staphylococcus phage P282 Staphylococcus phage StauST398-4 Staphylococcus phage phiN315 Staphylococcus phage phi7247PVL Staphylococcus phage phiSa2wa_st22 Staphylococcus virus 77 Staphylococcus phage P954
S9	Major capsid protein	IYDRNSDTLDGLPVVNLK	Staphylococcus aureus Staphylococcus argenteus	Staphylococcus virus 85 Staphylococcus phage SP5 Staphylococcus virus phiETA2 Staphylococcus phage phiNM2 Staphylococcus virus SAP26 Staphylococcus phage SA12 Staphylococcus virus Baq Sau1
S11 and S20	Phage repressor, Cro/CI family	ELAEAIGVSQPTVSNWIQQTK	Staphylococcus aureus Staphylococcus argenteus Staphylococcus sciuri	Staphylococcus virus IPLA35 Staphylococcus phage SMSAP5 Staphylococcus phage vB_SauS_phi2
S11 and S20	Phage repressor, Cro/CI family	IQQLADYFNVPK	Staphylococcus aureus Staphylococcus sciuri Staphylococcus pseudintermedius Staphylococcus devriesei Staphylococcus warneri Staphylococcus capitis Staphylococcus argenteus	Staphylococcus phage SMSAP5 Staphylococcus phage vB_SauS_phi2 Staphylococcus virus IPLA35
S12 S10 and S14	Complement inhibitor	IYNEIDEALKSK	Staphylococcus aureus, Enterobacter sp. IF2SW-B1 Klebsiella pneumoniae	Staphylococcus phage 13 Staphylococcus phage phiNM3 Staphylococcus phage StauST398-1
S20	Major capsid protein	VSYTLDDDDFITDVETAK	Staphylococcus aureus Staphylococcus haemolyticus Staphylococcus saprophyticus Staphylococcus warneri Staphylococcus argenteus Streptococcus pneumoniae Staphylococcus sciuri	Staphylococcus phage phiSa2wa_st72 Staphylococcus phage tp310-2 Staphylococcus phage phiSa2wa_st121mssa Staphylococcus phage vB_SauS_phi2 Staphylococcus phage StauST398-2 Staphylococcus virus 3a Staphylococcus phage LH1 Staphylococcus phage phiSa2wa_st30 Staphylococcus virus phi12 Staphylococcus virus phiSLT Staphylococcus phage vB_SauS_JS02 Staphylococcus phage R4 Staphylococcus phage vB_SauS_fPfSau02 Staphylococcus phage SA137ruMSSAST121PVL
S20	Phage protein (DUF2479 domain)	SIINGKLDSQWTVPNEHK	Staphylococcus aureus	Staphylococcus phage DW2 Staphylococcus virus IPLA88
S18	N-acetylmuramoyl-L-alanine amidase	KEAGNYTVANVK	Bacilli, Staphylococcus argenteus Staphylococcus aureus Staphylococcus sp. HMSC34H10	Staphylococcus phage tp310-1 Staphylococcus phage tp310-2 Staphylococcus phage phi2958PVL Staphylococcus phage PVL Staphylococcus phage SA137ruMSSAST121PVL Staphylococcus virus IPLA35
S4	Phage protein NrdI	VETFLENETNQNNLIAVM* SSGNRNWGTNFAIAGDTISK	Staphylococcus haemolyticus Staphylococcus hominis Staphylococcus aureus Staphylococcus aureus subsp. aureus Z172	
S12	Complement inhibitor	IYNEIDEALK	Staphylococcus. Aureus Klebsiella pneumoniae Enterobacter sp. IF2SW-B1	Staphylococcus phage StauST398-1 Staphylococcus virus 13
S10	Complement inhibitor	IYNEIDEALKSKY	Staphylococcus. aureus Klebsiella pneumoniae Enterobacter sp. IF2SW-B2	Staphylococcus phage StauST398-1 Staphylococcus virus 13
S10	DDE-type integrase/transposase/recombinase	PC*PALM*NKRNSIATIHR	Staphylococcus aureus	
S9	DNA primase phage-associated	LLHHFYNPENTTALSF NDLNDKFKPANLQGKLVNIAD	Staphylococcus aureus, Staphylococcus haemolyticus Staphylococcus capiti, Staphylococcus epidermidis Staphylococcus warneri Staphylococcus sp. HMSC077D08 Corynebacterium propinquum, Staphylococcus sp. U Staphylococcus lugdunensis Staphylococcus sp. HMSC077B09	Uncultured Caudovirales Phage

Table 1. Cont.

Strain	Protein	Peptide	Bacteria with 100% Homology Based on the NCBI Protein Database	Phages with 100% Homology Based on the NCBI Protein Database
S2	Phage repressor, Cro/CI family	AAHLEGELTDDEWQR	Staphylococcus haemolyticus Staphylococcus warneri Staphylococcus agnetis, Staphylococcus chromogenes Staphylococcus haemolyticus Staphylococcus sp. 58-22 Staphylococcus capitis Staphylococcus pasteuri Bacillales Staphylococcus chromogenes Staphylococcus agnetis Escherichia coli, Staphylococcus aureus 08-02906 Staphylococcus aureus VET0383R, Staphylococcus aureus VET0098R Staphylococcus aureus M1487 Staphylococcus aureus, Staphylococcus aureus A6300 Staphylococcus aureus subsp. aureus str. Newman Staphylococcus aureus subsp. aureus WBG10049, Staphylococcus aureus A9635, Staphylococcus aureus subsp. aureus MN8	Staphylococcus virus 71 Staphylococcus phage phiSa2wa_st1 Staphylococcus phage phiSa2wa_st5 Staphylococcus phage Henu2 Staphylococcus phage ROSA Staphylococcus phage phi7401PVL
S2	Phage repressor, Cro/CI family	VLDYADYIR	Staphylococcus aureus Staphylococcus epidermidis Staphylococcus warneri Staphylococcus agnetis Staphylococcus warneri Staphylococcus chromogenes, staphylococcus spp. Staphylococcus schleiferi Staphylococcus simulans Staphylococcus haemolyticus, Staphylococcus pettenkoferi Staphylococcus lugdunensis Escherichia coli	Staphylococcus virus 71 Staphylococcus phage phiSa2wa_st1 Staphylococcus phage phiSa2wa_st5 Staphylococcus phage Henu2 Staphylococcus phage ROSA Staphylococcus phage phi7401PVL
S9	DNA-binding protein	SLDNM*SLK	Striga asiática Staphylococcus aureus subsp. aureus 112808A Staphylococcus aureus A8819 Staphylococcus argenteus Staphylococcus spp. Pseudomonas aeruginosa Flectobacillus sp. BAB-3569 Eoetvoesia caeni Arabidopsis thaliana, Coxiellaceae bacterium, Clostridia bacterium	Staphylococcus phage vB_SauS_phi2
S19	DUF2479, Phage tail fiber, BppU family phage baseplate upper protein	HAGYVRC*KLF	Staphylococcus aureus, Staphylococcus sp. HMSC055H07 Staphylococcus argenteus, Staphylococcus sp. KY49P Staphylococcus sp. HMSC035F11 Pseudomonas aeruginosa Escherichia coli	Staphylococcus phage SA97 Staphylococcus virus 55 uncultured Caudovirales phage Staphylococcus virus 85 Staphylococcus virus 80 Staphylococcus virus phiETA3 Staphylococcus virus phiETA2 Staphylococcus phage 55-2 Staphylococcus phage B166 Staphylococcus phage B236 Staphylococcus virus SAP26 Staphylococcus virus 88 Staphylococcus virus phiETA Staphylococcus virus 11 Staphylococcus phage SP5 Staphylococcus virus 69 Staphylococcus phage ROSA Staphylococcus phage TEM123 Staphylococcus virus 92 Staphylococcus phage StauST398-1 Staphylococcus virus phiNM2 Staphylococcus virus phiNM1 Staphylococcus virus 29 Staphylococcus phage vB_SauS-SAP27 Staphylococcus virus 80alpha Staphylococcus phage HSA84 Staphylococcus virus phiMR11 Staphylococcus phage SAP33 Staphylococcus phage 3MRA
S12	Phage protein (DUF4393 domain)	NSPIDLNSTEISLNNLER	Staphylococcus aureus Staphylococcus spp. Staphylococcus argenteus	Staphylococcus phage StauST398-1
S12	Phage protein (DUF669 domain)	MNFNLNLQGAQELGN	Staphylococcus capitis Staphylococcus epidermidis Staphylococcus caprae Staphylococcus devriesei Staphylococcus warneri	Staphylococcus virus phiMR11
S10	GNAT family N-acetyltransferase	IINYARQNNYESLLTSIVSNNIGAK	Staphylococcus aureus Staphylococcus aureus subsp. anaerobius Staphylococcus aureus subsp. aureus Mu50 Staphylococcus hominis Escherichia coli	

Table 1. Cont.

Strain	Protein	Peptide	Bacteria with 100% Homology Based on the NCBI Protein Database	Phages with 100% Homology Based on the NCBI Protein Database
S5	Holin, phage phi LC3 family	SQDSNLTPELSTKAPK	Staphylococcus aureus	Staphylococcus phage HSA84 Staphylococcus phage SP5
S6	ImmA/IrrE family metallo-endopeptidase	EKAKIFGDFDMNDSGVY DEENSTIIYNPLDSITR	Staphylococcus aureus subsp. aureus H19 Staphylococcus aureus Staphylococcus aureus subsp. aureus Staphylococcus aureus subsp. aureus 21204	
S16	Involved in the expression of fibrinogen-binding protein phage-associated	ESINANTYINQNLEK	Staphylococcus aureus	
S16	Involved in the expression of fibrinogen-binding protein phage-associated	VAVLSTPLVTSFESK	Staphylococcus aureus	
S17	N-6 DNA methylase; N6_Mtase domain-containing protein	KDGEILFDAIDIYLRNK	Staphylococcus aureus	Staphylococcus phage phi-42
S4	Phage DNA-binding protein	GDM*FVVITIM*MQQIK	Staphylococcus aureus Staphylococcus warneri	
S9	Phage terminase	KLYIIEEYVKQGM	Staphylococcus aureus Staphylococcus argenteus Staphylococcus sp. HMSC58E11 Allobacillus sp. SKP4-8	Staphylococcus virus Baq_Sau1 Staphylococcus virus phiETA2 Staphylococcus virus 69 Staphylococcus virus 11 Staphylococcus virus 80alpha
S14	Integrase	M*PVYKDGNTGKWYFSI	Staphylococcus aureus	Staphylococcus phage B166 Staphylococcus virus phiMR25 Staphylococcus virus 88
S4	Phage repressor	ISKVQQLADYFNVPK	Staphylococcus aureus, Staphylococcus chromogenes Staphylococcus hyicus	Staphylococcus virus 80
S13	Toxin Phage protein; Pathogenicity island protein	NLDGVWLGDLILIKRGLSDR	Staphylococcus aureus, Staphylococcus sp. HMSC58E11, Staphylococcus argenteus, Escherichia coli	Staphylococcus phage phiSa2wa_st80 Staphylococcus phage 3MRA Staphylococcus phage phiSa2wa_st5
S16	Toxin Phage protein; Pathogenicity island protein	SDREKAGILFEELAHNK	Staphylococcus aureus Escherichia coli Staphylococcus argenteus Staphylococcus sp. HMSC58E11	Staphylococcus phage 3MRA Staphylococcus phage phiSa2wa_st5 Staphylococcus phage phiSa2wa_st80 Staphylococcus phage phiJB Staphylococcus phage phi7401PVL
S6	PBSX family phage terminase	QADNTYVHHSTYLNNP FISKQFIQEAESAKQR	Staphylococcus spp.	
S11	PBSX family phage terminase	QGVSHLFKVTKSPM*R	Staphylococcus aureus Staphylococcus lentus Staphylococcus sciuri	
S20	Phage-related cell wall hydrolase; Peptidase C51; CHAP domain-	EVPNEPDYIVIDVC*EDYSASK	Staphylococcus argenteus Staphylococcus sp. HMSC36F05	Staphylococcus virus IPLA88 Staphylococcus virus phiNM2 Staphylococcus phage SAP40 Staphylococcus phage phi 53 Staphylococcus virus phiNM4 Staphylococcus phage SA12 Staphylococcus virus 69 Staphylococcus phage SA97 Staphylococcus phage TEM123 Staphylococcus virus 11 Staphylococcus virus phiMR25 Staphylococcus virus 53 Staphylococcus phage SAP33
S5	Phage antirepressor Ant	QDWLAM*EVLPAIR	Staphylococcus aureus, Staphylococcus simulans Staphylococcus argenteus Staphylococcus pseudintermedius	Staphylococcus phage SA75 Staphylococcus phage SA13
S11	Phage capsid protein	M*AEETNSNVTEETEVNE	Staphylococcus, aureus Staphylococcus spp.	
S4	Phage encoded lipoprotein	IHDKELDDPSEEESKLTQEEENSI	Staphylococcus aureus, Staphylococcus capitis, Staphylococcus epidermidis, Staphylococcus cohnii, Staphylococcus haemolyticus	Staphylococcus phage SPbeta-like
S2	Phage head morphogenesis protein	KDVQRIVSHVT	Staphylococcus aureus Staphylococcus argenteus	
S9	YhgE/Pip, Phage infection protein	LNEYM*PNIEKLLN VASNDIPAQFPK	Staphylococcusaureus, Staphylococcus haemolyticus Staphylococcus sp. HMSC34C02	
S14	Minor structural protein	KTTSEALKEVLSDT	Staphylococcus aureus	
S4	Phage portal protein	EPKPVDATGADDPLKPDDRM* ITNFHANLVDQKVSY	Staphylococcus aureus	

Table 1. Cont.

Strain	Protein	Peptide	Bacteria with 100% Homology Based on the NCBI Protein Database	Phages with 100% Homology Based on the NCBI Protein Database
S5	Phage protein	VHISEFKYPLYM*DFLGTKGELE	Staphylococcus aureus Staphylococcus haemolyticus	
S15	Phage protein	MSHNALTTGIGIGAGAG	Staphylococcus aureus	
S2	Phage protein	EITDGEISSVLTM*M*	Staphylococcus aureus, Staphylococcus hominis Staphylococcus epidermidis	
S20	Phage recombination protein Bet	KSSTTYEVNGETVK	Staphylococcus aureus, Staphylococcus sciuri	
S2	Phage resistance protein	ESVDTGEITANTTRTVK	Staphylococcus aureus Staphylococcus fleurettii Staphylococcus pasteuri Staphylococcus epidermidis Staphylococcus warneri Staphylococcus schleiferi Escherichia coli	
S13	Tail tape measure protein	GM*PTGTNVYAVKGGIADK	Staphylococcus aureus, Staphylococcus saprophyticus, Staphylococcus pseudoxylosus	Staphylococcus phage phiSa2wa_st5 Staphylococcus phage phi3A Staphylococcus phage SH-St 15,644 Staphylococcus virus 3a
S3	Tail tape measure protein	VQHPGKLVNKVM*SGLNINFGGGANATAK	Staphylococcus aureus	
S4	Tail tape measure protein	QM*MEGLSGVMDLAAVSGEDLG AVSDIVTDGLTAFGLKAKDSG	Staphylococcus aureus	
S2	Tail tape measure protein	AEEAGVTVKQL	Staphylococcus aureus Staphylococcus cohnii Staphylococcus sp. HMSC061H04 Staphylococcus hominis Staphylococcus capitis Staphylococcus cohnii Staphylococcus sp. HMSC061H04 Staphylococcus sp. HMSC067G10 Staphylococcus Staphylococcus haemolyticus Enterococcus faecium Staphylococcus epidermidis Staphylococcus sp. HMSC067G10 Staphylococcus haemolyticus Enterococcus faecium Staphylococcus epidermidis	Staphylococcus phage SPbeta-like
S10	Phage repressor, Cro/CI family	QKNVLNYANEQLDEQNKV	Staphylococcus aureus, Bacilli, Staphylococcus hyicus Staphylococcus epidermidis	Staphylococcus virus phiNM2 Staphylococcus virus 53 Staphylococcus virus 80alpha
S13	Phage protein	KSNVEAFSNAVK	Staphylococcus aureus	Staphylococcus virus 80alpha Staphylococcus virus phiNM1 Staphylococcus virus phiNM2
S11	Phage protein	PYHDLSDERIM*EELKK	Staphylococcus aureus Staphylococcus argenteus taphylococcus schweitzeri	Staphylococcus virus phiETA2 Staphylococcus phage P630 Staphylococcus virus SAP26 Staphylococcus phage B236 Staphylococcus virus 88 Staphylococcus prophage phiPV83
S4	Minor structural protein	LNDNISNINTIV	Pseudomonas aeruginosa E. coli Pararheinheimera mesophila Staphylococcus pseudintermedius Staphylococcus epidermidis, Staphylococcus sp. KY49P Staphylococcus argenteus Staphylococcus schleiferi Staphylococcus hyicus Staphylococcus sp. HMSC063H12 Staphylococcus aureus	Staphylococcus virus 77 Staphylococcus phage P630 Staphylococcus phage SA780ruMSSAST101 Staphylococcus phage phiSa119 Staphylococcus phage phiN315 Staphylococcus phage SA7 Staphylococcus phage JS01 Staphylococcus phage StauST398-4 Staphylococcus virus 13 Staphylococcus phage 23MRA Staphylococcus virus 108PVL Staphylococcus phage phiBU01 Staphylococcus phage PVL Staphylococcus phage tp310-1 Staphylococcus phage P954 Staphylococcus phage SA345ruMSSAST8 Staphylococcus phage phiNM3 Staphylococcus virus 77 Staphylococcus phage phiSa2wa_st22 Staphylococcus phage SA1014ruMSSAST7 Staphylococcus phage P282 Staphylococcus prophage phiPV83 Staphylococcus phage 3 AJ-2017 Staphylococcus phage SAP090B Staphylococcus phage IME1346_01 Staphylococcus phage phi5967PVL Staphylococcus phage P1105 Staphylococcus phage IME1361_01

Table 1. Cont.

Strain	Protein	Peptide	Bacteria with 100% Homology Based on the NCBI Protein Database	Phages with 100% Homology Based on the NCBI Protein Database
S9	PhiETA ORF58-like protein	GMVASMQMQVVQVNVLTM*ELAQQNAMLTQQLTELK	Staphylococcus aureus	
S4	Phage portal protein	TEQLPRLEML	Staphylococcus aureus, Staphylococcus sp. HMSC063A07, Staphylococcus lugdunensis, Staphylococcus sp. HMSC068D08, Staphylococcus sp. HMSC069E09	
S4	Prophage, terminase	KDRYSSVSY	Staphylococcus aureus, Staphylococcus delphini, Staphylococcus pseudintermedius, Staphylococcus agnetis, Staphylococcus epidermidis, Staphylococcus hominis, Staphylococcus haemolyticus, Paenibacillus sophorae	Staphylococcus phage SPbeta-like
S4	Prophage tail domain; Peptidase	VLEM*IFLGEDPK	Staphylococcus aureus E. coli Bacilli	Staphylococcus phage phi7401PVL Staphylococcus phage phiSa2wa_st121mssa Staphylococcus virus 3a Staphylococcus virus phiSLT Staphylococcus phage tp310-2 Staphylococcus phage SA137ruMSSAST121PVL Staphylococcus phage phiSa2wa_st5 Staphylococcus phage phiSa2wa_st1 Staphylococcus phage SH-St 15644 Staphylococcus phage phi2958PVL Staphylococcus virus IPLA35 Staphylococcus phage P240 Staphylococcus phage vB_SauS_JS02 Staphylococcus virus 42e Staphylococcus virus phi12 Staphylococcus phage phiSa2wa_st72 Staphylococcus phage vB_SauS_fPfSau02 Staphylococcus phage phiSa2wa_st30 Staphylococcus phage vB_SauS_phi2 Staphylococcus phage StauST398-2
S15	Site-specific integrase	VEELEDSEIHKK	Staphylococcus aureus, Staphylococcus epidermidis Staphylococcus haemolyticus Staphylococcus condimenti Staphylococcus sp. HMSC035D11 Staphylococcus warneri	uncultured Caudovirales phage Sequence ID: ASN72447.1
S13	Site-specific integrase	KEAGSIINLHTINNALKSAC*R	Staphylococcus aureus Staphylococcus sp.	
S6	Site-specific integrase	YLNRNFVFTNHK	Staphylococcus aureus, Staphylococcus argenteus Staphylococcus cohini Staphylococcus lugdunensis Staphylococcus caeli Staphylococcus sp. 47.1	
S9	Terminase large subunit	KAMIKASPK	Staphylococcusaureus Escherichia coli Staphylococcus sp. HMSC74F04 Staphylococcus sp. HMSC055H07 Cutibacterium acnes Staphylococcus warneri Brevibacillus laterosporus Bacillus cihuensis Paenibacillus larvae	Staphylococcus phage vB_SauS_JS02 Staphylococcus phage Staphylococcus phage phiSa2wa_st5 Staphylococcus phage LH1 Staphylococcus phage phiSa2wa_st1 Staphylococcus phage phiSa2wa_st121mssa Staphylococcus virus IPLA35 Staphylococcus phage tp310-2 Staphylococcus virus phiSLT Staphylococcus phage StauST398-2 Staphylococcus phage vB_SauS_phi2 Staphylococcus virus phi12 Staphylococcus phage SMSAP5 Staphylococcus phage phi2958PVL Staphylococcus virus 3a Staphylococcus phage YMC/09/04/R1988
S20	Phage repressor, Cro/CI family	RIQQLADYFNVPK	Staphylococcus aureus Staphylococcus pettenkoferi Staphylococcus pettenkoferi Staphylococcus capitis Staphylococcus devriesei	Staphylococcus phage vB_SauS_phi2 Staphylococcus virus IPLA35

Table 1. Cont.

Strain	Protein	Peptide	Bacteria with 100% Homology Based on the NCBI Protein Database	Phages with 100% Homology Based on the NCBI Protein Database
S4	Transposase B from transposon Tn554 O	WDRRNLPLPDDK	Staphylococcus aureus, Staphylococcuspettenkoferi Staphylococcushominis, Quasibacillus thermotolerans Staphylococcaceae Staphylococcusvitulinus Streptococcus suis Staphylococcusfelis Salinicoccus roseus Staphylococcus epidermidis Staphylococcus lentus Staphylococcus warneri Staphylococcus epidermidis Staphylococcus chromogenes Staphylococcus sp. HMSC058E01 Enterococcus faecium Staphylococcus epidermidis VCU065 Staphylococcus cohnii Negativicoccus succinicivorans Eubacteriaceae bacterium Staphylococcus Enterococcus faecium Enterococcus Staphylococcus fleurettii Staphylococcus sp. 47.1 Bacilli Staphylococcus sp. SKL71207 Lactobacillales	
S13	Uncharacterized phage protein	C*VSGIAGGAVTGGTTLGLAGAG	Staphylococcus aureus Staphylococcus argenteus Staphylococcus schweitzeri Staphylococcus schweitzeri Staphylococcus hyicus Staphylococcus agnetis	
S13	Uncharacterized phage protein	DIITVYC*PENGTATDEY	Staphylococcus aureus	
S20	Uncharacterized phage protein	QTDVPSWVPM*VLR	Staphylococcusaureus Staphylococcus sp. HMSC74F04 Bacilli Staphylococcus Staphylococcus argenteus Staphylococcus sp. HMSC063H12	
S12	Uncharacterized phage protein	IIINHDEIDLL	Staphylococcus aureus Staphylococcus epidermidis Staphylococcus hominis Staphylococcus haemolyticus Staphylococcus sp. HMSC067G10 Staphylococcus haemolyticus Staphylococcus epidermidis Staphylococcus petrasii Staphylococcus capitis	Staphylococcus phage SPbeta-like
S14	Uncharacterized phage protein	TSIELITGFTK	Staphylococcus aureus, Staphylococcus sciuri, Staphylococcus schweitzeri, Staphylococcus spp.	Staphylococcus phage phi879, Staphylococcus phage phi575, Staphylococcus phage PVL, Staphylococcus prophage phiPV83, Staphylococcus phage SA45ruMSSAST97
S3	Uncharacterized phage protein	EFRNKLNELGADK	Staphylococcusaureus, Streptococcus pneumoniae, Terrabacteria group	Staphylococcus phage phi7401PVL, Staphylococcus phage tp310-2, Staphylococcus phage vB_SauS_phi2, Staphylococcus virus IPLA35, Staphylococcus phage phiSa2wa_st30, Staphylococcus virus 47, Staphylococcus virus 3a
S3	Phage repressor, Cro/CI family	HLEEVDIR	Staphylococcusaureus, Paxillus involutus ATCC 200175, Brassica cretica, Staphylococcus epidermidis, Staphylococcus spp., Enterobacter hormaechei	
S4	YhgE/Pip; Phage infection protein	APQSTSVKK	Staphylococcusaureus, Staphylococcusschweitzeri, Staphylococcus sp.	
S4	YhgE/Pip Phage infection protein	ALNFAADDVPAQFPK	S. aureus, Staphylococcus sp. HMSC36A10, Staphylococcus sp. HMSC34H10, Pseudomonas aeruginosa, E. coli	

3.2. Phage Peptides Determined from the Analyzed S. aureus Strains

For strains S2 and S3, six and three phage peptides were determined, respectively. For strain S4, seventeen phage peptides were determined, and three phage peptides were determined for strain S5. For strains S6 and S7, three and one phage peptides were determined, respectively. Moreover, for strains S8 and S9, two phage peptides and seven phage peptides were determined. For strains S10 and S11, five and three phage peptides were determined, respectively. For strains S12 and S13, five phage peptides and six phage peptides were determined, respectively. For strains S14 and S15, four and two phage peptides were determined, respectively. For strain S16, three phage peptides were determined, and one phage peptide was determined for strain S17. For strains S18 and S19, one phage peptide each was determined. Finally, for strain S20, seven phage peptides were determined.

A large number of phage peptides from structural proteins were identified (Table 1). Peptides from proteins such as the major capsid protein, major tail protein, minor structural protein, phage head morphogenesis protein, tail tape measure protein and phage tail fiber protein were determined. Moreover, different phage peptides from the major capsid protein and tail protein were determined (Table 1). Identifying these phage peptides is reasonable, as the major capsid protein and major tail protein are the most abundant proteins in mature virions [6].

There are a large number of uncharacterized protein sequences in databases, and more than 20% of all protein domains are annotated as "domains of unknown function" (DUFs). Several uncharacterized phage proteins and DUFs from *Staphylococcus* bacteriophages were identified for the analyzed strains (Table 1) [43,44].

Different peptides from repressor-type Cro/CI were determined. For strains S11 and S20 (both potential enterotoxin C producers), the same phage peptides of repressor-type Cro/CI were identified (Table 1). CI and Cro are encoded in the lysogeny module of lambdoid bacteriophages, particularly λ bacteriophages. Together, CII and CIII (that are formed through the anti-terminator role of protein N) act as an inducer that favors the first expression of the *cI* gene from the appropriate promoter; if the CI repressor predominates, the phage remains in the lysogenic state, but if the Cro predominates, the phage transitions into the lytic cycle, helped by the late Q regulator. The xenobiotic XRE regulator is extended in bacteria and has similarity to the Croλ repressor, exhibiting a helix-turn-helix (HTH) conformation [45]. Peptides of the CI/Cro-repressor types are usually named XRE family proteins in the NCBI database for bacteria.

Three phage peptides of the complement inhibitor were identified (Table 1). Staphylococcal complement inhibitors are involved in the evasion of human phagocytosis by blocking C3 convertases, and a study reported that complement inhibitor genes were also found in *staphylococcal* phages [46]. Another autolysin was determined in the present results, an N-acetylmuramoyl-L-alanine amidase that plays a role in bacterial adherence to eukaryotic cells [19]. The phage protein NrdI, which is a type of ribonucleotide reductase (RNR), was also identified. Several peptides of transposases, integrases and terminases were identified along with a DNA primase phage associated protein and a DNA phage binding protein. Moreover, peptides of other proteins, such as GNAT family N-acetyltransferase, holin, peptidase, methylase, anti-repressor protein (Ant), phage-resistant protein, phage-encoded lipoprotein, phage infection protein, phage portal protein, toxin phage proteins associated with pathogenicity islands and a protein involved in fibrinogen-binding proteins, were identified. A PBSX family phage terminase peptide was determined, and this protein is involved in double-stranded DNA binding, DNA packaging and endonuclease and ATPase activities [47].

As shown in Table 1, the vast majority of phage-specific peptides are not specific to *S. aureus* and can be found in other species of *Staphylococcus*. As an exception, the same peptides, such as peptide LLHALPTGNDSGGDKLLPK from a major capsid protein, were also found in *Streptococcus pneumoniae*, and peptide AYINITGLGFAK from a major tail protein was also found in *Pararheinheimera mesophila*; whether these examples represent direct recombinations between bacteria belonging to different families or whether phage-mediated recombination occurs remains to be elucidated. Furthermore, as mentioned before, eighteen identified peptides were very specific for *S. aureus* based on the NCBI database (see Figure S1).

3.3. Staphylococcus spp. Phage Genome Comparisons and Their Relatedness

A phylogenomic tree of *Staphylococcus* spp. phages from the NCBI database (accession numbers in Table S2 in Supplemental Data 2) with 100% similarity to those found in this study was built (Figure 2). The phages identified in this study were classified in the order *Caudovirales* and the family *Siphoviridae*. Many of these bacteriophages were classified into the genera *Phietavirus*, *Biseptimavirus*, *Triavirus* phieta-like virus, SPbeta-like virus and unclassified genera. Genomes of well-known phages of the families *Siphoviridae*, *Myoviridae* and *Podoviridae*, such as phage Lambda, T4 and T7, respectively, were added for comparison purposes. The genome analysis showed three well-defined clusters that mainly divided the phylogenomic tree into different phage genera (*Phietavirus*, *Biseptimavirus* and *Triavirus*). Two principal branches separated Clusters A, B and C from D. Cluster A was formed by *Staphylococcus Phietavirus*, two phieta-like viruses and two unclassified *Staphylococcus* phages. Cluster B was formed by *Staphylococcus* phages classified as *Biseptimavirus* and by one unclassified *Staphylococcus* phage. Cluster C was formed by enterobacterial bacteriophages and one SPbeta-like virus. Finally, cluster D was formed by *Triavirus Staphylococcus* phages and two unclassified *Staphylococcus* phages. To the best of our knowledge, this is the first time that phages from mastitis-causing staphylococci were grouped in a phylogenomic tree.

Specific peptides were found in related *Staphylococcus* spp. phages (Table 2) located closely in the phylogenomic tree (Figure 2). Peptides HAGYVRC*KLF and MPVYKDGNTGKWYFSI were found in phages of cluster A. Furthermore, peptides IYDRNSDTLDGLPVVNLK, QKNVLNYANEQLDEQNKV, EVPNEPDYIVIDVC*EDYSASK, KSNVEAFSNAVK and KLYIIEEYVKQGM were found in *Staphylococcus* phages of the A.1 subbranch in cluster A. Additionally, peptide AVAELLKEINR was found in phages of the A.2 branch. The peptide AYINITGLGFAK was found in phages of cluster B.1, and TSIELITGFTK was found in phages of cluster B.2. Peptides VSYTLDDDDFITDVETAK and LLHALPTGNDSGGD-KLLPK, which belong to the phage major capsid protein, were found in the same 14 *Staphylococcus* phages of cluster D. Peptides ELAEAIGVSQPTVSNWIQQTK and IQQLA-DYFNVPK, which belong to the phage-repressor Cro/CI family of proteins, were found in the same bacteriophages of cluster D. Moreover, peptides LYVGVFNPEATK, RVSYTLD-DDDFITDVETAKELKL LYVGVFNPEATK, VLEMIFLGEDPK, KAMIKASPK, EFRNKL-NELGADK and GMPTGTNVYAVKGGIADK were also found in phages of cluster D. Peptides IHDKELDDPSEEESKLTQEEENSI, IIINHDEIDLL, KDRYSSVSY and AEEAGVTVKQL are specific to *Staphylococcus* phage SPbeta-like.

Table 2. Phage biomarker peptides that belong to bacteriophages and phylogenomic tree clusters. Relationships between specific phage biomarker peptides and phylogenomic tree clusters.

Protein	Peptide	Phages	Cluster Located
Major capsid protein	VSYTLDDDDFITDVETAK	*Staphylococcus* phage phiSa2wa_st72 *Staphylococcus* phage tp310-2 *Staphylococcus* phage phiSa2wa_st121mssa *Staphylococcus* phage vB_SauS_phi2 *Staphylococcus* phage StauST398-2 *Staphylococcus* virus 3a *Staphylococcus* phage LH1 *Staphylococcus* phage phiSa2wa_st30 *Staphylococcus* virus phi12 *Staphylococcus* virus phiSLT *Staphylococcus* phage vB_SauS_JS02 *Staphylococcus* phage R4 *Staphylococcus* phage vB_SauS_fPfSau02 *Staphylococcus* phage SA137ruMSSAST121PVL	Cluster D

Table 2. Cont.

Protein	Peptide	Phages	Cluster Located
Major capsid protein	LLHALPTGNDSGGDKLLPK	Staphylococcus phage phiSa2wa_st72 Staphylococcus phage phiSa2wa_st121mssa Staphylococcus phage vB_SauS_phi2 Staphylococcus phage StauST398-2 Staphylococcus phage LH1 Staphylococcus phage phiSa2wa_st30 Staphylococcus virus phi12 Staphylococcus virus 3ª Staphylococcus virus phiSLT Staphylococcus phage tp310-2 Staphylococcus phage vB_SauS_JS02 Staphylococcus phage R4 Staphylococcus phage vB_SauS_fPfSau02 Staphylococcus phage SA137ruMSSAST121PVL	Cluster D
Major capsid protein	RVSYTLDDDDFITDVETAKELKL	Staphylococcus phage LH1 Staphylococcus phage StauST398-2 Staphylococcus phage vB_SauS_phi2 Staphylococcus phage R4	Cluster D
Major tail protein	LYVGVFNPEATK	Staphylococcus phage vB_SauS_ phi2 Staphylococcus virus phi12 Staphylococcus virus phiSLT Staphylococcus phage R4 Staphylococcus phage vB_SauS_JS02 Staphylococcus phage SH-St 15644 Staphylococcus virus 3a Staphylococcus phage P240	Cluster D
Phage repressor, Cro/CI family	ELAEAIGVSQPTVSNWIQQTK	Staphylococcus virus IPLA35 Staphylococcus phage SMSAP5 Staphylococcus phage vB_SauS_phi2	Cluster D
Phage repressor, Cro/CI family	IQQLADYFNVPK	Staphylococcus virus IPLA35 Staphylococcus phage SMSAP5 Staphylococcus phage vB_SauS_phi2	Cluster D
Major tail protein	AYINITGLGFAK	Staphylococcus phage phiNM3 Staphylococcus phage StauST398-4 Staphylococcus phage P282 Staphylococcus phage phiN315 Staphylococcus phage phi7247PVL Staphylococcus phage phiSa2wa_st22 Staphylococcus virus 77 Staphylococcus phage P954	Cluster B.1
Major capsid protein	IYDRNSDTLDGLPVVNLK	Staphylococcus virus 85 Staphylococcus phage SP5 Staphylococcus virus phiETA2 Staphylococcus phage phiNM Staphylococcus virus SAP26 Staphylococcus phage SA12 Staphylococcus virus Baq Sau1	Cluster A.1
Uncharacterized phage protein	AVAELLKEINR	Staphylococcus virus 71 Staphylococcus virus 55 Staphylococcus virus 88	Cluster A.2
DUF2479, Phage tail fiber, BppU family phage baseplate upper protein	HAGYVRCKLF	Staphylococcus phage SA97 Staphylococcus virus 55 uncultured Caudovirales phage Staphylococcus virus 85 Staphylococcus virus 80 Staphylococcus virus phiETA3 Staphylococcus virus phiETA2 Staphylococcus phage 55-2 Staphylococcus phage B166 Staphylococcus phage B236 Staphylococcus virus SAP26 Staphylococcus virus 88 Staphylococcus virus phiETA Staphylococcus virus 11 Staphylococcus phage SP5 Staphylococcus virus 69 Staphylococcus phage ROSA Staphylococcus phage TEM123 Staphylococcus virus 92 Staphylococcus phage StauST398-1 Staphylococcus virus phiNM2 Staphylococcus virus phiNM1 Staphylococcus virus 29 Staphylococcus phage vB_SauS-SAP27 Staphylococcus virus 80alpha Staphylococcus phage HSA84 Staphylococcus virus phiMR11 Staphylococcus phage SAP33 Staphylococcus phage 3MRA	Cluster A
Phage terminase	KLYIIEEYVKQGM	Staphylococcus virus Baq_Sau1 Staphylococcus virus phiETA2 Staphylococcus virus 69 Staphylococcus virus 11 Staphylococcus virus 80alpha	Cluster A.1
Phage-related cell wall hydrolase; Peptidase C51; CHAP domain-	EVPNEPDYIVIDVC*EDYSASK	Staphylococcus virus IPLA88 Staphylococcus virus phiNM2 Staphylococcus phage SAP40 Staphylococcus phage phi 53 Staphylococcus virus phiNM4 Staphylococcus phage SA12 Staphylococcus virus 69 Staphylococcus phage SA97 Staphylococcus phage TEM123 Staphylococcus virus 11 Staphylococcus virus phiMR25 Staphylococcus virus 53 Staphylococcus phage SAP33	Cluster A.1

Table 2. Cont.

Protein	Peptide	Phages	Cluster Located
Prophage_tail domain-; Peptidase	VLEM*IFLGEDPK	Staphylococcus phage phi7401PVL Staphylococcus phage phiSa2wa_st121mssa Staphylococcus virus 3a Staphylococcus virus phiSLT Staphylococcus phage tp310-2 Staphylococcus phage SA137ruMSSAST121PVL Staphylococcus phage phiSa2wa_st5 Staphylococcus phage phiSa2wa_st1 Staphylococcus phage SH-St 15644 Staphylococcus phage phi2958PVL Staphylococcus virus IPLA35 Staphylococcus phage P240 Staphylococcus phage vB_SauS_JS02 Staphylococcus virus 42e Staphylococcus virus phi12 Staphylococcus phage phiSa2wa_st72 Staphylococcus phage vB_SauS_fPfSau02 Staphylococcus phage phiSa2wa_st30 Staphylococcus phage vB_SauS_phi2 Staphylococcus phage StauST398-2	Cluster D
Terminase large subunit	KAM*IKASPK	Staphylococcus phage vB_SauS_JS02 Staphylococcus phage Staphylococcus phage phiSa2wa_st5 Staphylococcus phage LH1 Staphylococcus phage phiSa2wa_st1 Staphylococcus phage phiSa2wa_st121mssa Staphylococcus virus IPLA35 Staphylococcus phage tp310-2 Staphylococcus virus phiSLT Staphylococcus phage StauST398-2 Staphylococcus phage vB_SauS_phi2 Staphylococcus virus phi12 Staphylococcus phage SMSAP5 Staphylococcus phage phi2958PVL Staphylococcus virus 3a Staphylococcus phage YMC/09/04/R1988	Cluster D
Uncharacterized phage protein	TSIELITGFTK	Staphylococcus phage phi879, Staphylococcus phage phi575, Staphylococcus phage PVL, Staphylococcus prophage phiPV83, Staphylococcus phage SA45ruMSSAST97	Cluster B2
Uncharacterized phage protein	EFRNKLNELGADK	Staphylococcus phage phi7401PVL, Staphylococcus phage tp310-2, Staphylococcus phage vB_SauS_phi2, Staphylococcus virus IPLA35, Staphylococcus phage phiSa2wa_st30, Staphylococcus virus 47, Staphylococcus virus 3a	Cluster D
Phage protein	KSNVEAFSNAVK	Staphylococcus virus 80alpha Staphylococcus virus phiNM1 Staphylococcus virus phiNM2	Cluster A.1
Phage repressor, Cro/CI family	QKNVLNYANEQLDEQNKV	Staphylococcus virus phiNM2 Staphylococcus virus 53 Staphylococcus virus 80alpha	Cluster A.1
Tail tape measure protein	GM*PTGTNVYAVKGGIADK	Staphylococcus phage phiSa2wa_st5 Staphylococcus phage phi3A Staphylococcus phage SH-St 15,644 Staphylococcus virus 3a	Cluster D
integrase	M*PVYKDGNTGKWYFSI	Staphylococcus phage B166 Staphylococcus virus phiMR25 Staphylococcus virus 88	Cluster A

In addition, a correlation relating bacterial species for each cluster with all peptides found in the bacteriophages with 100% similarity was found. The results showed that clustered phages were related to specific species of *Staphylococcus*. All studied phages were found to be related to *S. aureus*; however, most of them were also found to be related to additional *Staphylococcus* species. *S. argenteus* was found to be related in all clusters of the phylogenomic tree. Cluster A phage peptides were found to be mainly related to *S. simiae*. However, different *Staphylococcus* species (*S. xylosus*, *S. muscae*, *S. haemolyticus*, *S. simiae*, *S. sciuri*, *S. pseudintermedius*, *S. devriesei*, *S. warneri* and *S. capitis*) were found to be related to phages of cluster D.

3.4. Identification of Peptides of Virulence Factors

In this work, 405 peptides from *S. aureus* were determined to be related to virulence factors (Excel dataset Supplemental Data). Among these peptides, proteins such as staphopain, beta-lactamase, elastin-binding protein peptides and a multidrug ATP-binding cassette (ABC) transporter were identified.

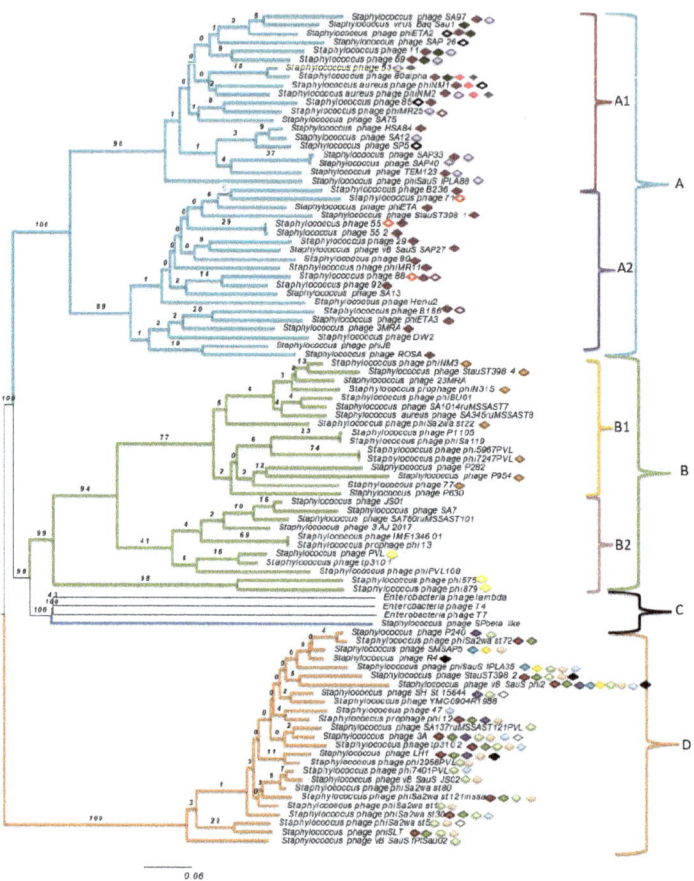

Figure 2. Phylogenomic tree generated by the Virus Classification and Tree Building Online Resource (VICTOR) using the complete genomic sequences of the determined *Staphylococcus* spp. phages. The access numbers of the determined phage genomes are shown in Table S2 in Supplemental Data 2. Genomes of the *lambda* (NC_001416.1), *T4* (NC_000866.4) and *T7* (NC_001604.1) phages were added for comparison purposes. The VICTOR phylogenetic tree construction was based on an intergenic distance analysis with the GBDP tool (Genome BLAST Distance Phylogeny). The significance of each branch is indicated by a pseudo-bootstrap value calculated as a percentage for 1000 subsets. Bar, 20 nt (nucleotides) substitutions per 100 nt. Clusters are represented by different colors: light blue, cluster A, red, cluster A.1, purple, cluster A.2, light green, cluster B, yellow, cluster B.1, pink, cluster B.2, black, cluster C and orange, cluster D. Specific cluster peptides are represented by different color forms: ◆, yellow-filled diamond IQQLADYFNVPK (cluster A-specific), ◆, brown-filled diamond HAGYVRC*KLF (cluster A-specific), ◇, black-outlined diamond IYDRNSDTLDGLPVVNLK (cluster A.1-specific), ◇, red=outlined diamond AVAELLKEINR (cluster A.2-specific), ◆, pink-filled diamond KSNVEAFSNAVK (cluster A.1), ◆, gray-filled diamond QKN-VLNYANEQLDEQNKV (cluster A.1), ◇, brown-outlined diamond MPVYKDGNTGKWYFSI (cluster A-specific), ◆, dark gray-filled diamond KLYIIEEYVKQGM (cluster A.1-specific), ◇, purple-outlined diamond EVPNEPDYIVIDVC*EDYSASK (cluster A.1-specific), ◆, orange-filled diamond AYINITGLGFAK (cluster B.1-specific), ◇, yellow-outlined diamond TSIELIT-GFTK (cluster B.2-specific), ◆, red-filled diamond VSYTLDDDDFITDVETAK (cluster D-specific), ◆, green-filled diamond LLHALPTGNDSGGDKLLPK (cluster D-specific), ◆, black-filled diamond RVSYTLDDDDFITDVETAKELKL (cluster D-specific), ◆, purple-filled diamond LYVGVFNPEATK (cluster D-specific), ◆, blue-filled diamond ELAEAIGVSQPTVSNWIQQTK (cluster D-specific); ◆, light green-filled diamond VLEMIFLGEDPK (cluster D-specific), ◇, orange-outlined diamond KAMIKASPK (cluster D-specific) and ◇, gray-outlined diamond GMPTGTNVYAVKGGIADK (cluster D-specific).

4. Discussion

LC-MS/MS-based methods for bacteriophage identification offer several advantages compared with other approaches, since bacteriophages can be directly identified with this method without using genomic tools, which provides a new strategy for drawing the appropriate conclusions. In addition, the method proposed here may be applied for further analyses without the requirement of growing bacteria, since the samples can be collected directly from foodstuffs. The study of noninduced prophages provides a fast analysis and can detect specific temperate phage proteins produced by *S. aureus* while integrated in the bacterial genome or by phages that are infecting the bacteria. Both cases provide the identification of specific *S. aureus* species or strains—in this case, an *S. aureus* mastitis producer. In the proteomic repository of the 20 different *S. aureus* strains analyzed, 79 peptides from staphylococcal bacteriophages were identified. Among them, eighteen of these phage peptides were *S. aureus*-specific. As bacteriophages are host-specific, these putative diagnostic peptides could be good diagnostic biomarkers for the detection and characterization of *S. aureus* and *S. aureus* phages.

The results show that a given specific peptide is present in closely related phages (Table 2). These bacteriophage peptides can be used as specific markers to establish *S. aureus* bacteriophage relationships (Figure 2). Additionally, phages that show the same peptides and are specific to *Staphylococcus* spp. are located close to one another in the phylogenomic tree, suggesting that a link does exist between phage phylogeny and bacteriophages that can infect the same bacterial species.

The study shown here exemplifies how phylogenomic trees based on the genome analysis provide useful information, and the study corroborates previous investigations, which suggested that viral genomic or subgenomic region analyses provide the best tool for reconstructing viral evolutionary histories [48]. Nevertheless, the lack of knowledge of the phage genomic content [49] makes a phage analysis more difficult. The first priority must be the contribution of new large amounts of data for phages infecting bacteria [12].

In addition, there is an urgent need for novel therapies to treat and prevent mastitis [50]. Bacteriophage therapy is an alternative to the antibiotic treatment of bovine mastitis [51], with a high specificity and a low probability for bacterial resistance development [52]. Many studies have demonstrated the effectiveness of bacteriophages in a variety of animal models to fight several mastitis-causing pathogenic bacteria. Some studies have shown how virulent phages such as SPW and SA phages are active against bovine mastitis-associated *S. aureus*. Moreover, SAJK-IND and MSP phages have specific lytic activity against several strains of *S. aureus* isolated from mastitis milk samples [53]. Indeed, mouse-induced mastitis models decreased their bacterial counts after treatment with a vBSM-A1 and vBSP-A2 phage cocktail [54]. Finally, several temperate phage mixtures have been shown to be more effective than using a single temperate phage for inhibiting *S. aureus*. According to the data obtained for the different models of mastitis, phage therapy using bacteriophages in this study can be considered an innovative alternative to antibiotics for the treatment of mastitis caused by *S. aureus*.

Finally, the proteomic analysis by LC-ESI-MS/MS performed in this study provides relevant insights into the search for potential phage origin diagnostic peptide biomarkers for mastitis-causing *S. aureus*. In addition, this method may be useful for searching peptide biomarkers for the identification and characterization of mastitis-causing species and for finding new *S. aureus* phages useful as possible therapies for mastitis.

Supplementary Materials: The following are available online at https://www.mdpi.com/article/10.3390/foods10040799/s1: Figure S1: MS/MS spectrums for *S. aureus*-specific peptide biomarkers. The corresponding peptides were tested for specificity using the BLASTp algorithm. Excel Dataset Supplemental Data 1: Complete nonredundant peptide dataset. Supplemental Data 2: Table S1: *Staphylococcus aureus* (SA) strains used in this study. Table S2: Linage, authors and accession number of studied bacteriophages [55–88].

Author Contributions: A.G.A. wrote the manuscript; A.G.A., K.B., T.G.V., P.C.-M., B.C., J.B.-V. J.-L.R.R. and M.C. conceptualized, revised and corrected the paper. P.C.-M. and M.C. co-supervised the work. M.C. and P.C.-M. got the funding. All authors listed have made a substantial, direct and intellectual contribution to the work and approved the work for publication.

Funding: This work received financial support from the Xunta de Galicia and the European Union (European Social Fund-ESF), from the Spanish Ministry of Economy and Competitivity Project AGL 2.013-48.244-R and from the European Regional Development Fund (ERDF) (2007–2013). This work was also supported by the GAIN-Xunta de Galicia Project (IN607D 2017/01) and the Spanish AEI/EU-FEDER PID2019-103845RB-C21 project. Mónica Carrera was supported by the Ramón y Cajal contract (Ministry of Science and Innovation of Spain).

Institutional Review Board Statement: Not applicable.

Informed Consent Statement: Not applicable.

Data Availability Statement: All relevant data are included in the article. The mass spectrometric data were deposited into the public database PRIDE (Proteomics Identification Database), with the dataset identifier PXD023530.

Acknowledgments: The mass spectrometry proteomics data were deposited into the ProteomeXchange Consortium via the PRIDE [89] partner repository with the dataset identifier PXD023530.

Conflicts of Interest: The authors declare no conflicts of interest.

References

1. Böhme, K.; Morandi, S.; Cremonesi, P.; Fernández No, I.C.; Barros-Velázquez, J.; Castiglioni, B.; Brasca, M.; Cañas, B.; Calo-Mata, P. Characterization of *Staphylococcus aureus* strains isolated from Italian dairy products by MALDI-TOF mass fingerprinting. *Electrophoresis* **2012**, *33*, 2355–2364. [CrossRef]
2. Forsman, P.; Tilsala-Timisjärvi, A.; Alatossava, T. Identification of staphylococcal and streptococcal causes of bovine mastitis using 16S-23S rRNA spacer regions. *Microbiology* **1997**, *143*, 3491–3500. [CrossRef]
3. Carrera, M.; Böhme, K.; Gallardo, J.M.; Barros-Velázquez, J.; Cañas, B.; Calo-Mata, P. Characterization of foodborne strains of *Staphylococcus aureus* by shotgun proteomics: Functional networks, virulence factors and species-specific peptide biomarkers. *Front. Microbiol.* **2017**, *8*, 2458. [CrossRef] [PubMed]
4. Rainard, P.; Foucras, G.; Fitzgerald, J.R.; Watts, J.L.; Koop, G.; Middleton, J.R. Knowledge gaps and research priorities in *Staphylococcus aureus* mastitis control. *Transbound. Emerg. Dis.* **2018**, *65*, 149–165. [CrossRef]
5. Abril, A.G.; Villa, T.G.; Barros-Velázquez, J.; Cañas, B.; Sánchez-Pérez, A.; Calo-Mata, P.; Carrera, M. *Staphylococcus aureus* exotoxins and their detection in the dairy industry and mastitis. *Toxins* **2020**, *12*, 537. [CrossRef]
6. Xia, G.; Wolz, C. Phages of *Staphylococcus aureus* and their impact on host evolution. *Infect. Genet. Evol.* **2014**, *21*, 593–601. [CrossRef] [PubMed]
7. Fortier, L.C.; Sekulovic, O. Importance of prophages to evolution and virulence of bacterial pathogens. *Virulence* **2013**, *4*, 354–365. [CrossRef] [PubMed]
8. Menouni, R.; Hutinet, G.; Petit, M.A.; Ansaldi, M. Bacterial genome remodeling through bacteriophage recombination. *FEMS Microbiol. Lett.* **2015**, *362*, 1–10. [CrossRef]
9. Deghorain, M.; Van Melderen, L. The staphylococci phages family: An overview. *Viruses* **2012**, *4*, 3316–3335. [CrossRef]
10. Feiner, R.; Argov, T.; Rabinovich, L.; Sigal, N.; Borovok, I.; Herskovits, A.A. A new perspective on lysogeny: Prophages as active regulatory switches of bacteria. *Nat. Rev. Microbiol.* **2015**, *13*, 641–650. [CrossRef]
11. Penadés, J.R.; Chen, J.; Quiles-Puchalt, N.; Carpena, N.; Novick, R.P. Bacteriophage-mediated spread of bacterial virulence genes. *Curr. Opin. Microbiol.* **2015**, *23*, 171–178. [CrossRef]
12. Brüssow, H.; Desiere, F. Comparative phage genomics and the evolution of *Siphoviridae*: Insights from dairy phages. *Mol. Microbiol.* **2001**, *39*, 213–222. [CrossRef]
13. Canchaya, C.; Fournous, G.; Brüssow, H. The impact of prophages on bacterial chromosomes. *Mol. Microbiol.* **2004**, *53*, 9–18. [CrossRef]
14. Uchiyama, J.; Taniguchi, M.; Kurokawa, K.; Takemura-Uchiyama, I.; Ujihara, T.; Shimakura, H.; Sakaguchi, Y.; Murakami, H.; Sakaguchi, M.; Matsuzaki, S. Adsorption of *Staphylococcus* viruses S13' and S24-1 on *Staphylococcus aureus* strains with different glycosidic linkage patterns of wall teichoic acids. *J. Gen. Virol.* **2017**, *98*, 2171–2180. [CrossRef]
15. Moon, B.Y.; Park, J.Y.; Hwang, S.Y.; Robinson, D.A.; Thomas, J.C.; Fitzgerald, J.R.; Park, Y.H.; Seo, K.S. Phage-mediated horizontal transfer of a *Staphylococcus aureus* virulence-associated genomic island. *Sci. Rep.* **2015**, *5*, 9784. [CrossRef]
16. Koskella, B.; Brockhurst, M.A. Bacteria-phage coevolution as a driver of ecological and evolutionary processes in microbial communities. *FEMS Microbiol. Rev.* **2014**, *38*, 916–931. [CrossRef]
17. Chakravorty, S.; Helb, D.; Burday, M.; Connell, N.; Alland, D. A detailed analysis of 16S ribosomal RNA gene segments for the diagnosis of pathogenic bacteria. *J. Microbiol. Methods* **2007**, *69*, 330–339. [CrossRef]

8. Ivnitski, D.; Abdel-hamid, I.; Atanasov, P.; Wilkins, E. Biosensors for detection of pathogenic bacteria. *Biosens. Bioelectron.* **1999**, *14*, 599–624. [CrossRef]
9. Abril, A.G.; Carrera, M.; Böhme, K.; Barros, J.; CANAS, B.; Rama, J.L.R.; Villa, T.G.; Calo-Mata, P. Characterization of bacteriophage peptides of pathogenic *Streptococcus* by LC-ESI-MS/MS: Bacteriophage phylogenomics and their relationship to their host. *Front. Microbiol.* **2020**, *11*, 1241. [CrossRef]
10. Gantzer, C.; Maul, A.; Audic, J.M.; Pharmacie, D. Detection of infectious enteroviruses, enterovirus genomes, somatic coliphages, and bacteroides fragilis phages in treated wastewater. *Appl. Environ. Microbiol.* **1998**, *64*, 4307–4312. [CrossRef]
11. Böhme, K.; Fernández-No, I.C.; Barros-Velázquez, J.; Gallardo, J.M.; Cañas, B.; Calo-Mata, P. Rapid species identification of seafood spoilage and pathogenic Gram-positive bacteria by MALDI-TOF mass fingerprinting. *Electrophoresis* **2011**, *32*, 2951–2965. [CrossRef] [PubMed]
12. Branquinho, R.; Sousa, C.; Lopes, J.; Pintado, M.E.; Peixe, L.V.; Osorio, H. Differentiation of *Bacillus pumilus* and *Bacillus safensis* using MALDI-TOF-MS. *PLoS ONE* **2014**, *9*, e110127. [CrossRef] [PubMed]
13. Lasch, P.; Beyer, W.; Nattermann, H.; Stämmler, M.; Siegbrecht, E.; Grunow, R.; Naumann, D. Identification of *Bacillus anthracis* by using matrix-assisted laser desorption ionization-time of flight mass spectrometry and artificial neural networks. *Appl. Environ. Microbiol.* **2009**, *75*, 7229–7242. [CrossRef] [PubMed]
14. Quintela-Baluja, M.; Böhme, K.; Fernández-No, I.C.; Alnakip, M.E.; Caamano, S.; Barros-Velázques, J.; Calo-mata, P. MALDI-TOF Mass Spectrometry, a rapid and reliable method for the identification of bacterial species in food-microbiology Laboratories. *Nov. Food Preserv. Microb. Assess. Tech.* **2014**, 353–385.
15. Craigie, J.; Yen, C.H. The Demonstration of Types of *B. typhosus* by means of preparations of type ii vi phage: I. Principles and Technique on JSTOR. *Can. J. Public Health* **1938**, *29*, 484–496.
16. Chanishvili, N. *Nanotechnology to Aid Chemical and Biological Defense*; Springer: Berlin/Heidelberg, Germany, 2015; pp. 17–33.
17. Lavigne, R.; Ceyssens, P.; Robben, J. Phage proteomics: Applications of mass spectrometry. In *Bacteriophages*; Humana Press: Totowa, NJ, USA, 2009; Volume 502, pp. 239–251.
18. Rees, J.C.; Voorhees, K.J. Simultaneous detection of two bacterial pathogens using bacteriophage amplification coupled with matrix-assisted laser desorption/ionization time-of-flight mass spectrometry. *Rapid Commun. Mass Spectrom.* **2005**, *19*, 2757–2761. [CrossRef]
19. Richter, Ł.; Janczuk-richter, M.; Niedzió, J.; Paczesny, J.; Ho, R. Recent advances in bacteriophage-based methods for bacteria detection. *Drug Discov. Today* **2018**, *23*, 448–455. [CrossRef]
20. Singh, A.; Poshtiban, S.; Evoy, S. Recent advances in bacteriophage based biosensors for food-borne pathogen detection. *Sensors* **2013**, *13*, 1763–1786. [CrossRef] [PubMed]
21. Calo-Mata, P.; Carrera, M.; Böhme, K.; Caamaño-Antelo, S.; Gallardo, J.M.; Barros-Velázquez, J.; Cañas, B. Novel Peptide Biomarker discovery for detection and identification of bacterial pathogens by LC-ESI-MS/MS. *J. Anal. Bioanal. Tech.* **2016**, *7*, 296.
22. Pfrunder, S.; Grossmann, J.; Hunziker, P.; Brunisholz, R.; Gekenidis, M.-T.; Drissner, D. *Bacillus cereus* group-type strain-specific diagnostic peptides. *J. Proteome Res.* **2016**, *15*, 3098–3107.
23. Serafim, V.; Ring, C.; Pantoja, L.; Shah, H.S.A. Rapid identification of *E. coli* bacteriophages using Mass Spectrometry. *J. Proteom. Enzymol.* **2017**, *6*, 1000130.
24. Morandi, S.; Brasca, M.; Lodi, R.; Cremonesi, P.; Castiglioni, B. Detection of classical enterotoxins and identification of enterotoxin genes in *Staphylococcus aureus* from milk and dairy products. *Vet. Microbiol.* **2007**, *124*, 66–72. [CrossRef]
25. Giebel, R.; Worden, C.; Rust, S.M.; Kleinheinz, G.T.; Robbins, M.; Sandrin, T.R. Microbial fingerprinting using matrix-assisted laser desorption ionization time-of-flight mass spectrometry (MALDI-TOF MS) applications and challenges. *Adv. Appl. Microbiol.* **2010**, *71*, 149–184.
26. Böhme, K.; Fernández-No, I.C.; Barros-Velázquez, J.; Gallardo, J.M.; Calo-Mata, P.; Cañas, B. Species differentiation of seafood spoilage and pathogenic gram-negative bacteria by MALDI-TOF mass fingerprinting. *J. Proteome Res.* **2010**, *9*, 3169–3183. [CrossRef]
27. Böhme, K.; Fernández-No, I.C.; Barros-Velázquez, J.; Gallardo, J.M.; Cañas, B.; Calo-Mata, P. Comparative analysis of protein extraction methods for the identification of seafood-borne pathogenic and spoilage bacteria by MALDI-TOF mass spectrometry. *Anal. Methods* **2010**, *2*, 1941. [CrossRef]
28. Carrera, M.; Cañas, B.; Gallardo, J.M. The sarcoplasmic fish proteome: Pathways, metabolic networks and potential bioactive peptides for nutritional inferences. *J. Proteomics* **2013**, *78*, 211–220. [CrossRef]
29. Käll, L.; Canterbury, J.D.; Weston, J.; Noble, W.S.; MacCoss, M.J. Semi-supervised learning for peptide identification from shotgun proteomics datasets. *Nat. Methods* **2007**, *4*, 923–925. [CrossRef]
30. Altschul, S.F.; Gish, W.; Miller, W.; Myers, E.W.; Lipman, D.J. Basic local alignment search tool. *J. Mol. Biol.* **1990**, *215*, 403–410. [CrossRef]
31. Meier-Kolthoff, J.P.; Göker, M. VICTOR: Genome-based phylogeny and classification of prokaryotic viruses. *Bioinformatics* **2017**, *33*, 3396–3404. [CrossRef]
32. Lucchini, S.; Desiere, F.; Brüssow, H. Similarly organized lysogeny modules in temperate *Siphoviridae* from low GC content gram-positive bacteria. *Virology* **1999**, *263*, 427–435. [CrossRef]
33. Bateman, A.; Coggill, P.; Finn, R.D. DUFs: Families in search of function. *Acta Crystallogr. Sect. F Struct. Biol. Cryst. Commun.* **2010**, *66*, 1148–1152. [CrossRef] [PubMed]

44. Goodacre, N.F.; Gerloff, D.L.; Uetz, P. protein domains of unknown function are essential in Bacteria. *MBio* **2014**, *5*, e00744-13 [CrossRef] [PubMed]
45. Durante-Rodríguez, G.; Mancheño, J.M.; Díaz, E.; Carmona, M. Refactoring the λ phage lytic/lysogenic decision with a synthetic regulator. *Microbiologyopen* **2016**, *5*, 575–581. [CrossRef]
46. Van Wamel, W.J.; Rooijakkers, S.H.; Ruyken, M.; van Kessel, K.P.; van Strijp, J.A. The Innate Immune Modulators staphylococcal complement inhibitor and chemotaxis inhibitory protein of *Staphylococcus aureus* are located on β the innate immune modulators staphylococcal complement inhibitor and chemotaxis inhibitory protein of *Staphylococcus*. *J. Bacteriol.* **2006**, *188*, 1310–1315. [PubMed]
47. Gual, A.; Camacho, A.G.; Alonso, J.C. Functional analysis of terminase large subunit, G2P, of *Bacillus subtilis* bacteriophage SPP1. *J. Biol. Chem.* **2000**, *275*, 35311–35319. [CrossRef] [PubMed]
48. Simmonds, P. Methods for virus classification and the challenge of incorporating metagenomic sequence data. *J. Gen. Virol.* **2015**, *96*, 1193–1206. [CrossRef] [PubMed]
49. Argov, T.; Azulay, G.; Pasechnek, A.; Stadnyuk, O.; Ran-Sapir, S.; Borovok, I.; Sigal, N.; Herskovits, A.A. Temperate bacteriophages as regulators of host behavior. *Curr. Opin. Microbiol.* **2017**, *38*, 81–87. [CrossRef] [PubMed]
50. Angelopoulou, A.; Warda, A.K.; Hill, C.; Ross, R.P. Non-antibiotic microbial solutions for bovine mastitis–live biotherapeutics, bacteriophage, and phage lysins. *Crit. Rev. Microbiol.* **2019**, *45*, 564–580. [CrossRef]
51. Lin, D.M.; Koskella, B.; Lin, H.C. Phage therapy: An alternative to antibiotics in the age of multi-drug resistance. *World J. Gastrointest. Pharmacol. Ther.* **2017**, *8*, 162. [CrossRef]
52. Dams, D.; Briers, Y. Enzybiotics: Enzyme-based antibacterials as therapeutics. In *Advances in Experimental Medicine and Biology*; Springer: New York, NY, USA, 2019; Volume 1148, pp. 233–253.
53. Ganaie, M.Y.; Qureshi, S.; Kashoo, Z.; Wani, S.A.; Hussain, M.I.; Kumar, R.; Maqbool, R.; Sikander, P.; Banday, M.S.; Malla, W.A.; et al. Isolation and characterization of two lytic bacteriophages against *Staphylococcus aureus* from India: Newer therapeutic agents against Bovine mastitis. *Vet. Res. Commun.* **2018**, *42*, 289–295. [CrossRef]
54. Geng, H.; Zou, W.; Zhang, M.; Xu, L.; Liu, F.; Li, X.; Wang, L.; Xu, Y. Evaluation of phage therapy in the treatment of *Staphylococcus aureus*-induced mastitis in mice. *Folia Microbiol.* **2019**, *65*, 339–351. [CrossRef]
55. Kwan, T.; Liu, J.; DuBow, M.; Gros, P.; Pelletier, J. The complete genomes and proteomes of 27 *Staphylococcus aureus* bacteriophages. *Proc. Natl. Acad. Sci. USA* **2005**, *102*, 5174–5179. [CrossRef]
56. Bae, T.; Baba, T.; Hiramatsu, K.; Schneewind, O. Prophages of *Staphylococcus aureus* Newman and their contribution to virulence. *Mol. Microbiol.* **2006**, *62*, 1035–1047. [CrossRef] [PubMed]
57. Kuroda, M.; Ohta, T.; Uchiyama, I.; Baba, T.; Yuzawa, H.; Kobayashi, I.; Cui, L.; Oguchi, A.; Aoki, K.; Nagai, Y.; et al. Whole genome sequencing of meticillin-resistant *Staphylococcus aureus*. *Lancet* **2001**, *357*, 1225–1240. [CrossRef]
58. Keary, R.; McAuliffe, O.; Ross, R.P.; Hill, C.; O'Mahony, J.; Coffey, A. Genome analysis of the staphylococcal temperate phage DW2 and functional studies on the endolysin and tail hydrolase. *Bacteriophage* **2014**, *4*, e28451. [CrossRef]
59. Van der Mee-Marquet, N.; Corvaglia, A.R.; Valentin, A.S.; Hernandez, D.; Bertrand, X.; Girard, M.; Kluytmans, J.; Donnio, P.Y.; Quentin, R.; François, P. Analysis of prophages harbored by the human-adapted subpopulation of *Staphylococcus aureus* CC398. *Infect. Genet. Evol.* **2013**, *18*, 299–308. [CrossRef]
60. García, P.; Martínez, B.; Obeso, J.M.; Lavigne, R.; Lurz, R.; Rodríguez, A. Functional genomic analysis of two *Staphylococcus aureus* phages isolated from the dairy environment. *Appl. Environ. Microbiol.* **2009**, *75*, 7663–7673. [CrossRef]
61. Yoon, H.; Yun, J.; Lim, J.A.; Roh, E.; Jung, K.S.; Chang, Y.; Ryu, S.; Heu, S. Characterization and genomic analysis of two *Staphylococcus aureus* bacteriophages isolated from poultry/livestock farms. *J. Gen. Virol.* **2013**, *94*, 2569–2576. [CrossRef]
62. Iandolo, J.J.; Worrell, V.; Groicher, K.H.; Qian, Y.; Tian, R.; Kenton, S.; Dorman, A.; Ji, H.; Lin, S.; Loh, P.; et al. Comparative analysis of the genomes of the temperate bacteriophages φ11, φ12 and φ13 of *Staphylococcus aureus* 8325. *Gene* **2002**, *289*, 109–118. [CrossRef]
63. Zhang, M.; Ito, T.; Li, S.; Jin, J.; Takeuchi, F.; Lauderdale, T.-L.Y.; Higashide, M.; Hiramatsu, K. Identification of the third type of PVL phage in ST59 methicillin-resistant *Staphylococcus aureus* (MRSA) strains. *FEMS Microbiol. Lett.* **2011**, *323*, 20–28. [CrossRef] [PubMed]
64. El Haddad, L.; Moineau, S. Characterization of a novel panton-valentine leukocidin (PVL)-encoding staphylococcal phage and its naturally PVL-lacking variant. *Appl. Environ. Microbiol.* **2013**, *79*, 2828–2832. [CrossRef] [PubMed]
65. Liu, J.; Dehbi, M.; Moeck, G.; Arhin, F.; Banda, P.; Bergeron, D.; Callejo, M.; Ferretti, V.; Ha, N.; Kwan, T.; et al. Antimicrobial drug discovery through bacteriophage genomics. *Nat. Biotechnol.* **2004**, *22*, 185–191. [CrossRef] [PubMed]
66. Kraushaar, B.; Hammerl, J.A.; Kienöl, M.; Heinig, M.L.; Sperling, N.; Thanh, M.D.; Reetz, J.; Jäckel, C.; Fetsch, A.; Hertwig, S. Acquisition of virulence factors in livestock-associated MRSA: Lysogenic conversion of CC398 strains by virulence gene-containing phages. *Sci. Rep.* **2017**, *7*, 1–13. [CrossRef]
67. Narita, S.; Kaneko, J.; Chiba, J.I.; Piémont, Y.; Jarraud, S.; Etienne, J.; Kamio, Y. Phage conversion of Panton-Valentine leukocidin in *Staphylococcus aureus*: Molecular analysis of a PVL-converting phage, φSLT. *Gene* **2001**, *268*, 195–206. [CrossRef]
68. Chang, Y.; Lee, J.H.; Shin, H.; Heu, S.; Ryu, S. Characterization and complete genome sequence analysis of *Staphylococcus aureus* bacteriophage SA12. *Virus Genes* **2013**, *47*, 389–393. [CrossRef]
69. Xiao, X.M.; Ito, T.; Kondo, Y.; Cho, M.; Yoshizawa, Y.; Kaneko, J.; Katai, A.; Higashiide, M.; Li, S.; Hiramatsu, K. Two different Panton-Valentine leukocidin phage lineages predominate in Japan. *J. Clin. Microbiol.* **2008**, *46*, 3246–3258.

Kaneko, J.; Kimura, T.; Narita, S.; Tomita, T.; Kamio, Y. Complete nucleotide sequence and molecular characterization of the temperate staphylococcal bacteriophage φPVL carrying Panton-Valentine leukocidin genes. *Gene* **1998**, *215*, 57–67. [CrossRef]

Mariem, B.J.J.; Ito, T.; Zhang, M.; Jin, J.; Li, S.; Ilhem, B.B.B.; Adnan, H.; Han, X.; Hiramatsu, K. Molecular characterization of methicillin-resistant Panton-valentine leukocidin positive *Staphylococcus aureus* clones disseminating in Tunisian hospitals and in the community. *BMC Microbiol.* **2013**, *13*, 2. [CrossRef] [PubMed]

Christie, G.E.; Matthews, A.M.; King, D.G.; Lane, K.D.; Olivarez, N.P.; Tallent, S.M.; Gill, S.R.; Novick, R.P. The complete genomes of *Staphylococcus aureus* bacteriophages 80 and 80α-Implications for the specificity of SaPI mobilization. *Virology* **2010**, *407*, 381–390. [CrossRef]

Frígols, B.; Quiles-Puchalt, N.; Mir-Sanchis, I.; Donderis, J.; Elena, S.F.; Buckling, A.; Novick, R.P.; Marina, A.; Penadés, J.R. Virus Satellites Drive Viral Evolution and Ecology. *PLoS Genet.* **2015**, *11*, e1005609. [CrossRef]

Botka, T.; Růžičková, V.; Konečná, H.; Pantůček, R.; Rychlík, I.; Zdráhal, Z.; Petráš, P.; Doškař, J. Complete genome analysis of two new bacteriophages isolated from impetigo strains of *Staphylococcus aureus*. *Virus Genes* **2015**, *51*, 122–131. [CrossRef]

Yamaguchi, T.; Hayashi, T.; Takami, H.; Nakasone, K.; Ohnishi, M.; Nakayama, K.; Yamada, S.; Komatsuzawa, H.; Sugai, M. Phage conversion of exfoliative toxin A production in *Staphylococcus aureus*. *Mol. Microbiol.* **2000**, *38*, 694–705. [CrossRef] [PubMed]

Santiago-Rodriguez, T.M.; Naidu, M.; Jones, M.B.; Ly, M.; Pride, D.T. Identification of staphylococcal phage with reduced transcription in human blood through transcriptome sequencing. *Front. Microbiol.* **2015**, *6*, 216. [CrossRef] [PubMed]

Matsuzaki, S.; Yasuda, M.; Nishikawa, H.; Kuroda, M.; Ujihara, T.; Shuin, T.; Shen, Y.; Jin, Z.; Fujimoto, S.; Nasimuzzaman, M.D.; et al. Experimental protection of mice against lethal *Staphylococcus aureus* infection by novel bacteriophage φMR11. *J. Infect. Dis.* **2003**, *187*, 613–624. [CrossRef] [PubMed]

Carroll, D.; Kehoe, M.A.; Cavanagh, D.; Coleman, D.C. Novel organization of the site-specific integration and excision recombination functions of the *Staphylococcus aureus* serotype F virulence-converting phages φ13 and φ42. *Mol. Microbiol.* **1995**, *16*, 877–893. [CrossRef] [PubMed]

Hoshiba, H.; Uchiyama, J.; Kato, S.I.; Ujihara, T.; Muraoka, A.; Daibata, M.; Wakiguchi, H.; Matsuzaki, S. Isolation and characterization of a novel *Staphylococcus aureus* bacteriophage, φMR25, and its therapeutic potential. *Arch. Virol.* **2010**, *155*, 545–552. [CrossRef] [PubMed]

Varga, M.; Pantůček, R.; Růžičková, V.; Doškař, J. Molecular characterization of a new efficiently transducing bacteriophage identified in meticillin-resistant *Staphylococcus aureus*. *J. Gen. Virol.* **2016**, *97*, 258–268. [CrossRef]

Pantůček, R.; Doškař, J.; Růžičková, V.; Kašpárek, P.; Oráčová, E.; Kvardová, V.; Rosypal, S. Identification of bacteriophage types and their carriage in *Staphylococcus aureus*. *Arch. Virol.* **2004**, *149*, 1689–1703. [CrossRef]

Chang, Y.; Shin, H.; Lee, J.-H.; Park, C.; Paik, S.-Y.; Ryu, S. Isolation and genome characterization of the virulent *Staphylococcus aureus* bacteriophage SA97. *Viruses* **2015**, *7*, 5225–5242. [CrossRef]

Zou, D.; Kaneko, J.; Narita, S.; Kamio, Y. Prophage, φpv83-pro, carrying panton-valentine leukocidin genes, on the *Staphylococcus aureus* p83 chromosome: Comparative analysis of the genome structures of φpv83-pro, φpvl, φ11, and other phages. *Biosci. Biotechnol. Biochem.* **2000**, *64*, 2631–2643. [CrossRef]

Utter, B.; Deutsch, D.R.; Schuch, R.; Winer, B.Y.; Verratti, K.; Bishop-Lilly, K.; Sozhamannan, S.; Fischetti, V.A. Beyond the Chromosome: The prevalence of unique extra-chromosomal bacteriophages with integrated virulence genes in pathogenic *Staphylococcus aureus*. *PLoS ONE* **2014**, *9*, e100502. [CrossRef]

Sanchini, A.; Del Grosso, M.; Villa, L.; Ammendolia, M.G.; Superti, F.; Monaco, M.; Pantosti, A. Typing of Panton-Valentine leukocidin-encoding phages carried by methicillin-susceptible and methicillin-resistant *Staphylococcus aureus* from Italy. *Clin. Microbiol. Infect.* **2014**, *20*, O840–O846. [CrossRef] [PubMed]

Jia, H.; Bai, Q.; Yang, Y.; Yao, H. Complete genome sequence of *Staphylococcus aureus* siphovirus phage JS01. *Genome Announc.* **2013**, *1*, 797–810. [CrossRef] [PubMed]

Jeon, J.; D'Souza, R.; Hong, S.K.; Lee, Y.; Yong, D.; Choi, J.; Lee, K.; Chong, Y. Complete genome sequence of the bacteriophage YMC/09/04/R1988 MRSA BP: A lytic phage from a methicillin-resistant *Staphylococcus aureus* isolate. *FEMS Microbiol. Lett.* **2014**, *359*, 144–146. [CrossRef] [PubMed]

Zeman, M.; Mašlaňová, I.; Indráková, A.; Šiborová, M.; Mikulášek, K.; Bendíčková, K.; Plevka, P.; Vrbovská, V.; Zdráhal, Z.; Doškař, J.; et al. *Staphylococcus sciuri* bacteriophages double-convert for staphylokinase and phospholipase, mediate interspecies plasmid transduction, and package mecA gene. *Sci. Rep.* **2017**, *7*, 46319. [CrossRef] [PubMed]

Perez-Riverol, Y.; Csordas, A.; Bai, J.; Bernal-Llinares, M.; Hewapathirana, S.; Kundu, D.J.; Inuganti, A.; Griss, J.; Mayer, G.; Eisenacher, M.; et al. The PRIDE database and related tools and resources in 2019: Improving support for quantification data. *Nucleic Acids Res.* **2019**, *47*, 442–450. [CrossRef]

Article

Validation of a MS Based Proteomics Method for Milk and Egg Quantification in Cookies at the Lowest VITAL Levels: An Alternative to the Use of Precautionary Labeling

Linda Monaci [1,*], **Elisabetta De Angelis** [1], **Rocco Guagnano** [1], **Aristide P. Ganci** [2], **Ignazio Garaguso** [3], **Alessandro Fiocchi** [4] **and Rosa Pilolli** [1]

1. Institute of Sciences of Food Production, CNR-ISPA, 70126 Bari, Italy; elisabetta.deangelis@ispa.cnr.it (E.D.A.); rocco.guagnano@ispa.cnr.it (R.G.); rosa.pilolli@ispa.cnr.it (R.P.)
2. PerkinElmer Italia S.p.A., Viale dell'Innovazione 3, 20126 Milano, Italy; aristide.ganci@perkinelmer.com
3. PerkinElmer LAS Germany GmbH, Ferdinand-Porsche-Ring 17, 63110 Rodgau, Germany; ignazio.garaguso@perkinelmer.com
4. Allergy Division, Bambino Gesù Children's Hospital, Istituti di Ricovero e Cura a Carattere Scientifico, 00165 Rome, Italy; Alessandro.fiocchi@allegriallergia.net
* Correspondence: linda.monaci@ispa.cnr.it

Received: 27 August 2020; Accepted: 12 October 2020; Published: 19 October 2020

Abstract: The prevalence of food allergy has increased over the last decades and consequently the food labeling policies have improved over the time in different countries to regulate allergen presence in foods. In particular, Reg 1169 in EU mandates the labelling of 14 allergens whenever intentionally added to foods, but the inadvertent contamination by allergens still remains an uncovered topic. In order to warn consumers on the risk of cross-contamination occurring in certain categories of foods, a precautionary allergen labelling system has been put in place by food industries on a voluntary basis. In order to reduce the overuse of precautionary allergen labelling (PAL), reference doses and action limits have been proposed by the Voluntary Incidental Trace Allergen Labelling VITAL project representing a guide in this jeopardizing scenario. Development of sensitive and reliable mass spectrometry methods are therefore of paramount importance in this regard to check the contamination levels in foods. In this paper we describe the development of a time-managed multiple reaction monitoring (MRM) method based on a triple quadrupole platform for milk and egg quantification in processed food. The method was in house validated and allowed to achieve levels of proteins lower than 0.2 mg of total milk and egg proteins, respectively, in cookies, challenging the doses recommended by VITAL. The method was finally applied to cookies labeled as milk and egg-free. This method could represent, in perspective, a promising tool to be implemented along the food chain to detect even tiny amounts of allergens contaminating food commodities.

Keywords: egg; milk; allergens; multiple reaction monitoring; mass spectrometry; reference doses; food; PAL

1. Introduction

The most recent epidemiology studies show the continuous increasing prevalence of food allergy worldwide and highlight global disparities of the incidence proportion, influenced by numerous genetic and environmental factors, as well as by gene–environment interactions [1,2]. The main treatment for sensitive individuals appears to be the lifelong avoidance of the offending foods [3]. In order to safeguard the health of sensitive consumers, European Commission Regulation No. 1169/2011 established the list of 14 allergenic ingredients (and by-products) whose presence must be indicated in the respective food labels whenever incorporated into foods. The list includes the following ingredients: milk, egg, cereals containing gluten, fish, crustacean, peanut, soy, tree nuts (hazelnut, almond, walnut, cashew, pecan nuts, Brazil nuts, pistachio, macadamia), sesame, lupin, mustard, celery, mussels, and sulphur dioxide (sulphite) [4]. However, current legislation does not address the unintentional occurrence of allergens due to cross-contamination along the entire food chain, neither established legal threshold levels for managing hidden allergens, posing a relevant health risk to allergic consumers [5]. To fill this gap, various countries have recently set own legal thresholds (e.g., Switzerland, Germany, Belgium, and Netherlands), lacking, however, harmonization among the different legal entities. In this frame, the European project ThRAll, funded by the European Food Safety Agency will actively contribute to the harmonization of MS-based methods by developing a prototype quantitative reference method for the multiple detection of food allergens in incurred food matrices [6].

Since 2007, in absence of official regulatory thresholds and facing the complexity of food allergen management, Australia and New Zealand developed the Voluntary Incidental Trace Allergen Labelling (VITAL) system to assist food producers in managing cross-contamination along the supply chains [7]. This system establishes eliciting doses (EDs) based on clinical studies for the protection of at least 95% (ED05) or 99% (ED01) of allergic people [7–9]. Recently, the version 3.0 of the VITAL program was released and for milk and egg proteins it set 0.2 mg total protein of allergenic ingredient as reference dose for action level 1, meaning that below this threshold no precautionary labelling statement is required, and 99% of the allergic population would safely consume the food. To comply with such threshold levels, reliable and sensitive methods are needed for the identification and quantification of allergenic contaminants.

So far, ELISA and PCR represent the techniques most commonly implemented across the laboratories for food allergen control. The limitations affecting these technologies such as cross-reactivity, low inter-assay reproducibility, missing multiplexing ability for ELISA, and the restrictions due to specificity for DNA based method have moved the attention towards LC–MS-based methods, representing a sequence-specific, protein-based approach [10–13]. Several multiplexing methods using multiple reaction monitoring (MRM) on low resolution mass spectrometers or alternative high-resolution based MS analysis have been reported and recently reviewed [12], all proving the sensitivity and reliability of an MS based analytical approach. Noteworthy, only a few of them were developed and validated on incurred food matrices [14–23].

The present work aims at evaluating the performance of a targeted multiple reaction monitoring (MRM) MS method using a last generation triple quadrupole mass spectrometer for the simultaneous detection of milk and egg allergens contamination in model bakery products, namely cookies. Synthetic peptides were used for method development and validation. In particular, the cookie reference material (RM) developed by MoniQA Association was used for the estimation of method recovery. This RM was specifically designed to performance evaluation of milk-detection methods and its production mimic as closely as possible the actual manufacturing process. Finally, the developed method was applied to the analysis of real samples to detect milk and egg traces in commercial cookies labelled as "milk and egg allergen free".

2. Materials and Methods

2.1. Chemicals and Materials

Solvents and reagents were purchased from Sigma–Aldrich (Milan, Italy) while Trypsin Gold Mass Spectrometry Grade was purchased from Promega (Milan, Italy). Ultrapure water was produced by a Millipore Milli-Q system (Millipore, Bedford, MA, USA) while formic acid (MS grade) was purchased from Fluka (Milan, Italy). Disposable desalting cartridges PD-10 were purchased from GE Healthcare Life Sciences (Milan, Italy) while syringe filters (0.45 µm of porosity in regenerated cellulose RC, and 5 µm of porosity in cellulose acetate CA) were purchased from Sartorius (Gottingem,

Germany). Sep-Pak C18 cartridges (50 mg, 1 mL) were obtained from Waters s.p.a. (Milan, Italy). Skim milk powder and whole egg powder were purchased by Sigma Aldrich (Milan, Italy).

For the preparation of matrix matched calibration curves, allergen-free and incurred cookies were produced at laboratory scale according to the recipe already described in a previous paper [19]. The incurred cookie was prepared at a high contamination level and diluted with blank cookie to match the final concentration required.

Cookie reference materials (RM) for milk allergen detection were purchased from MoniQA association (Güssing, Austria). The kit contains the following four samples: (i) a positive control consisting of characterized dried skim milk powder (SMP-MQA 092014) with validated protein content; (ii) a negative control gluten-free cookie (BLANK-MQA 082015), and two incurred materials (gluten free cookies) added with SMP at two concentration levels, (iii) low inclusion level (LOW-MQA 102016, concentration approx. 10 $mg_{allergenic\ ingredient}$/kg equivalent to 3.54 $mg_{milk\ protein}$/kg), (iv) high inclusion level (HIGH-MQA 082016, concentration approx. 50 $mg_{allergenic\ ingredient}$/kg equivalent to 17.7 $mg_{milk\ protein}$/kg). Ten different lots of blind commercial cookies labeled by the manufacturer as "prepared without adding of milk and eggs" were provided by Galbusera SpA (Cosio Valtellino, Sondrio, Italy).

2.2. Synthetic Peptides Standard Solutions

Native synthetic peptides (Table S1) were synthetized by GenScript (Piscataway, NJ, USA) and distributed by Twin Helix (Milan, Italy). Peptide purity was composed of between 90% and 99% as confirmed by HPLC analysis, while the respective mass was proved by MS analysis. Peptides for each allergen were received as lyophilized powder and reconstituted with 100 mM Bicarbonate Ammonium/Acetonitrile (80/20; v/v) to reach the concentration of 1 mg/mL. Reconstituted peptides were then aliquoted in a 0.5 mL tube and stored at −20 °C until use.

2.3. Sample Preparation Protocol

Firstly, allergen free and incurred cookie prepared at laboratory scale together with commercial cookies were ground mechanically and sifted with a 1-mm sieve. Conditions for total protein extraction, purification, and digestion were described elsewhere [19,20] with few modifications. In particular, the extraction buffer was replaced by Tris–HCl buffer 200 mM with Urea 7 M at pH = 9.2 and the resulting extract was filtered through 5 μm acetate cellulose membranes. Trypsin digestion was stopped after 14 h by acidification (HCl 6 M) and the final digest was centrifuged at 1800× g for 10 min before collecting the supernatant. Tryptic digest was then filtered through a 0.45 μm regenerated cellulose (RC) filter and 1 mL aliquot loaded on a C18 SPE column (previously conditioned with methanol and 50 mM ammonium bicarbonate) for a further purification step. C18-retained peptides were washed with 800 μL of 0.1% formic acid aqueous solution and eluted with 1.5 mL of methanol/water (90:10 v/v). The collected fraction was dried under gentle air stream and suspended in 100 μL of 0.1% formic acid in acetonitrile/water (90/10, v/v solution). Samples were finally filtered through a RC 0.45 μm syringe filter. The analytical workflow for sample preparation is schematized in Figure 1.

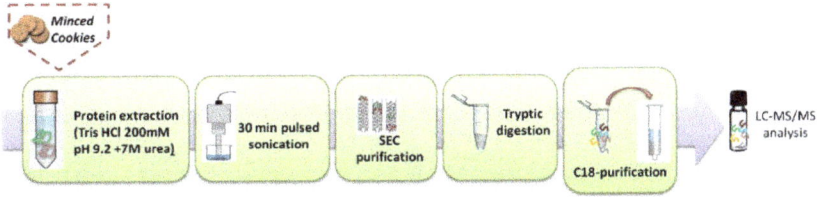

Figure 1. Experimental workflow for the simultaneous detection of milk and egg allergens in cookie samples.

2.4. Liquid Chromatography–Multiple Reaction Monitoring Analysis

LC–MRM analysis was performed on a PerkinElmer UHPLC LX50 System (PerkinElmer Inc., Waltham, MA, USA) coupled with a PerkinElmer QSight® 220 MS/MS (PerkinElmer Inc., Waltham, MA, USA) detector based on triple quadrupole mass analyzer. Peptide mixture (injection volume 10 µL) was separated on a Perkin Elmer Aqueous C18 Column (2.1 × 150 mm; 3 µm; 100 Å) (PerkinElmer Inc., Waltham, MA, USA). LC method parameters are detailed in Table S1 while MRM conditions are summarized in Table S2. MRM data were acquired in positive ion mode at unit resolution (0.7 ± 0.1 amu) in both Q1 and Q3. ESI source parameters were set as follows: drying gas (nitrogen): 120 (arbitrary units); HSID™ Temp: 250 °C; Nebulizer gas: 300 (arbitrary units); ion source T °C: 400. All instrument control, analysis, and data processing were performed using the Simplicity™ 3Q software platform v. 1.4 (PerkinElmer Inc., Waltham, MA, USA).

2.5. Performance Evaluation for In-House Method Validation

2.5.1. Sensitivity

A matrix matched calibration curve was prepared over the concentration range of 0.0125–0.25 µg/mL (four concentration levels) by spiking a defined amount of synthetic peptide stock solutions to tryptic digest of allergen-free cookie extract. All calibration points were filtered using 0.45 µm filters and then injected (10 µL) in duplicate on the column. Native synthetic peptide peak areas were acquired and by applying proper conversion factors (see Figure 2 for details) the reporting units were converted into total proteins of allergenic ingredient (µg/g). Main analytical criteria, such as sensitivity, repeatability/reproducibility, recovery, and processing effect, were evaluated according to these reporting units.

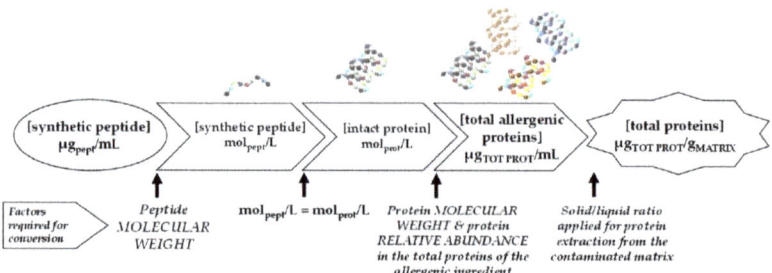

Figure 2. Flowchart calculation for the conversion of egg and milk synthetic peptides concentration ($\mu g_{peptide}/mL_{extract}$) into total protein concentration ($\mu g_{tot\ prot}/g_{matrix}$).

In order to evaluate any eventual effect of processing on the sensitivity of the method, matrix-matched calibration curves prepared by fortifying cookies with allergenic ingredients before processing (incurred samples) were built up for each milk and egg allergen marker selected. Specifically, five concentration levels were prepared in the range 10–300 $\mu g_{allergenic\ ingredient}/g_{matrix}$. As first level, a cookie incurred at 3000 $\mu g_{allergenic\ ingredient}/g_{matrix}$ was produced and then submitted to protein extraction and dilution with the blank extract to obtain the point at 300 $\mu g_{allergenic\ ingredient}/g_{matrix}$. Calibration points at lower concentrations were produced by progressive dilution of the highest level with blank cookies extract. All extracts were then submitted to SEC purification, tryptic digestion, and peptide purification on C18-SPE to be finally filtered on 0.45 µm filters and then injected (10 µL) in duplicate on HPLC/MS equipment. Peptide peak areas were acquired, and the reporting units were converted into total proteins of allergenic ingredient (µg/g) by assuming 35.39% and 48.05% of total protein content for milk and egg ingredients, respectively, in accordance with previous chemical characterization analysis performed on the allergic materials used for cookie production.

2.5.2. Precision

For method precision, a single contamination level at 100 $\mu g_{allergenic\ ingredient}/g_{matrix}$ was analyzed. Five analytical replicates were prepared and analyzed (intra-day repeatability). The same analyses were repeated over three different days and compared by one-way ANOVA test at 95% confidence level.

2.5.3. Trueness

Method recovery was evaluated only for milk by means of the validated RMs developed by MoniQA association. The blank sample provided with the kit was used to create a new matrix-matched calibration curve with synthetic peptides. The LOW and HIGH incurred samples were analyzed and the percent ratio between the measured and the validated concentration values defined the method recovery.

3. Results and Discussion

3.1. Optimization of LC–MS Instrumental Conditions

A sensitive method based on HPLC separation and mass spectrometry detection equipped with triple quadrupole analyzer for the simultaneous detection of milk and egg allergens in a model bakery product, namely cookie, was developed. The proteomic bottom-up approach was applied by detecting proteotypic peptides for monitoring food contamination by allergenic ingredients. Both milk and egg are widely investigated allergens and as such, a good consensus about the most reliable peptide markers has been achieved already by independent investigations [24]. The peptides that arose from tryptic digestion of αS1-casein, namely FFVAPFPEVFGK (FFV) and YLGYLEQLLR (YLG), and from β-lactoglobulin, namely TPEVDDEALEK (TPE) and VLVLDTDYK (VLV) were used for tracking milk, and peptides belonging to ovalbumin, ISQAVHAAHAEINEAGR (ISQ) and GGLEPINFQTAADQAR (GGL) and to vitellogenin-II namely NIPFAEYPTYK (NIP) and NIGELGVEK (NIG) were chosen for egg detection. All these markers have been already validated by previous works accomplished on bakery products [24–26].

In order to build up an analytical method for absolute quantitation, synthetic analogous of the aforementioned peptide sequences were purchased. Firstly, standard solutions of such peptides were prepared and injected in flow analysis for the optimization of instrumental parameters setting up the MRM detection on triple quadrupoles. For each peptide, the three most sensitive transitions were selected and collision energies, entrance voltages, and collision cell lens voltages were tuned to maximize the signal to noise ratio (Table S2). The chromatographic conditions for peptide separation were optimized and the best compromise between total running time and peak resolution was found.

In order to confirm the absence of interfering peaks from the matrix background, a blank cookie sample was prepared according to the sample preparation protocol described in Section 2.4 and added with synthetic peptides at fixed concentration. In Figure 3, a typical chromatogram acquired under the best separation conditions is presented and averaged peak retention times are reported.

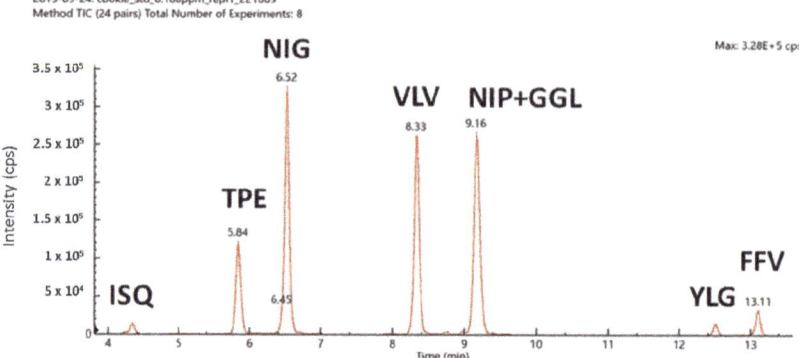

Figure 3. Typical chromatograms recorded for synthetic peptides in cookie matrix (total ion current, peptide concentration level 0.166 µg/mL).

3.2. Sensitivity and Matrix Effect

After optimizing the instrumental conditions, different aliquots of blank cookie samples were added with increasing concentration of milk and egg synthetic peptides in order to build-up matrix-matched calibration curves. In particular, four calibration points within the range of 0.125–0.25 µg/mL were prepared and the linear interpolation of resulting peak areas allowed evaluating the linearity range, and the sensitivity for each precursor/transition acquired.

One of the controversial aspects in food allergen detection has been the reporting unit of the contamination level. As well known the legislation refers to allergen labelling as whole ingredient, although clinical studies and potential threshold levels refer to the total protein content of the allergenic ingredient.

Specifically, protein is the hazard that causes allergic reactions, therefore analytical methods reporting contamination level as mg of total proteins would streamline the usability of the information retrieved also in light of the adherence to prescribed threshold levels and of the consistency of method sensitivity to reference doses of the VITAL Program. This issue represents an important bottleneck for mass spectrometric detection where for absolute quantitation, peptide-based calibration curves, and further conversions from the peptide units into total protein units, are required. Practically, in order to calculate final protein concentration, the peptide concentration in the digest volume (µg/mL) needs to be converted into total protein of the allergenic ingredient in matrix weight (µg/g). Until now, no international agreement about proper conversion factors has been achieved and only few examples from previous literature are to date available [15,23] on this regard. In this investigation, we applied a similar conversion scheme presented in Figure 2. Both milk and egg have been widely investigated in terms of protein composition, therefore the information available in the literature was used to retrieve proper conversion factors based on specific mathematical calculation and molar equivalence as schematized in Figure 2. Briefly, for each synthetic peptide, peptide concentration in the tryptic digest (reported as µg/mL) was first converted into molarity and then, assuming the complete release of each peptide from its parent protein, protein molarities were calculated. Afterwards, based on protein molecular weight and its relative abundance within the total proteins contained in food ingredient we calculated the total allergen proteins per mL of digest. Finally, as last conversion step, by taking into consideration the solid/liquid ratio used for sample extraction (1:20), we obtained the required reporting unit of µg of total protein of the allergenic ingredient per g of matrix. By following this approach all peptide reporting units were converted into $\mu g_{total\ protein}/g_{matrix}$ providing an analytical range between 1.3 and 680 µg/g depending of the specific marker. The new reporting units were integrated in the matrix-matched calibration curves and all method performance features were referred to them. The response linearity obtained in the matrix-matched calibration curve was very good for all the peptide markers monitored within the investigated range, with linear correlation coefficients at least better than 0.9859. Limit of detection (LOD) and quantification (LOQ) were calculated according to the interpolation parameters as 3-times and 10-times, respectively, the standard deviation of the line intercept divided by the slope. The careful evaluation of LOD/LOQ values for the detected transitions allowed to identify the best quantifier marker and its most sensitive transition as reported in Table 1.

These analyses allowed us to further evaluate the matrix effect on the peptides chosen. According to our results, very challenging LODs were achieved, as low as 0.1 and 3 $\mu g_{tot\ prot}/g_{matrix}$, respectively, and referred to FFV and TPE αS1-casein and β-lactoglobulin peptides for milk allergen. As for egg, LODs of 0.3 and 3 $\mu g_{tot\ prot}/g_{matrix}$ were found for ISQ (ovalbumin) and NIP (vitellogenin-2) peptides, respectively. LOQ values are also reported in Table 1. Noteworthy, the sensitivity provided by the peptides TPE and NIP was lower than the peptides FFV and ISQ, respectively; however, it is important to keep them in the analytical method as specific markers of whey and yolk proteins, notwithstanding their lower relevance from the allergological point of view, to encompass also risk of contamination from partial milk/egg based formulations. In Figure 4, it is shown a typical chromatogram obtained for the milk (FFV-m/z 692.9→991.4 and TPE-m/z 623.3→572.5) and egg allergens (ISQ-m/z 592.1→858.9 and NIP-m/z 671.8→557.9) and their relevant confirmative transitions in the cookie sample.

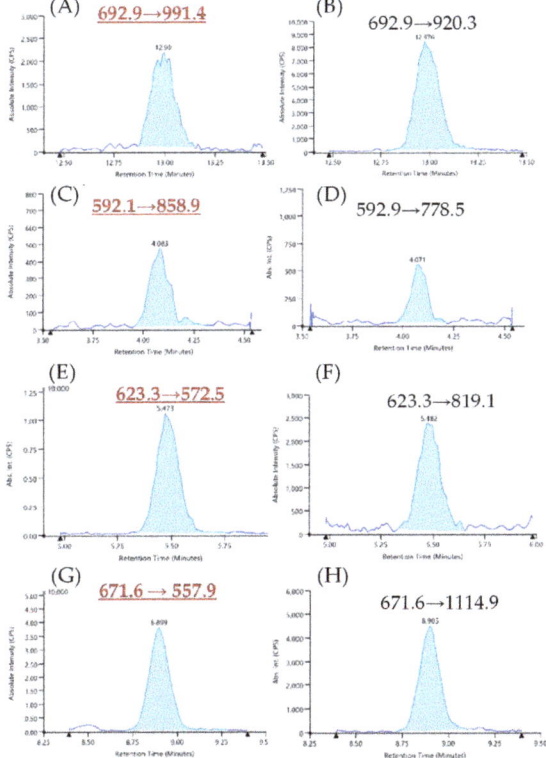

Figure 4. Typical chromatograms acquired for synthetic peptides in the cookie matrix: Extracted Ion Chromatogram XIC of quantifier transitions for most sensitive marker peptides of milk ((**A**), FFV = m/z 692.9→991.4; (**C**), TPE = m/z 623.3→572.5) along with their relevant qualifier transition ((**B**), FFV = m/z 692.9→920.3; (**D**), TPE = m/z 623.3→819.1) and egg, quantifier transitions ((**E**), ISQ = m/z 592.1→858.9; (**G**), NIP = m/z 671.6→557.9); qualifier transitions ((**F**), ISQ = m/z 592.1→778.5; (**H**), NIP = m/z 671.6→1114.9) at a level of 0.0125 μg/mL.

Table 1. Analytical features of the developed analytical method on the basis of synthetic peptides matrix-matched calibration curves.

Allergenic Ingredient: Protein	Marker	Quantifier Transition	LOD/LOQ (μg TOT PROT/g MATRIX)	R^2	CV% intra Day 1	CV% intra Day 2	CV% intra Day 4	CV% inter	Recovery LOW-MQA Material	Recovery HIGH-MQA Material
Milk: αS1-Casein	FFV	692.9→991.4	0.10/0.3	1.0000	1.9%	0.7%	6%	4%	57 ± 4%	50 ± 3%
Milk: β-Lactoglobulin	TPE	623.3→572.5	3/8	0.9992	8%	9%	10%	9%	-	-
Egg: Ovalbumin	ISQ	592.1→858.9	0.3/1.1	1.0000	2%	5%	1.7%	4%	-	-
Egg: Vitellogenin-2	NIP	671.8→557.9	3/9	1.0000	3%	3%	4%	6%	-	-

3.3. Sensitivity of the Method in Incurred Cookies and Compliance with the VITAL Reference Doses

As already mentioned, the VITAL grid was developed in 2007, aiming at providing a helpful management tool for food producers as well as to consumers. Although originally created by the Allergen Bureau of Australia and New Zealand, this system has been taken into consideration and used as reference values by numerous countries within the European Union until other official and harmonized limits will be available for the different allergenic foods. In particular, the VITAL Program provides a quantitative method for risk-assessment to evaluate the impact of allergen cross-contamination and to make decisions regarding proper precautionary allergen management and labeling. This approach allows not only safeguarding the health of allergic consumers, but also preserving the value of precautionary labeling as a risk management tool, avoiding its massive use also in very low-risk cases. The likelihood to develop an adverse reaction in allergic people depends on the total amount of allergenic proteins consumed during a meal, and on the level of sensitization of each individual. Therefore, the crucial point in the VITAL Program was to find a correlation between these two topics and define the maximum concentration level from accidental contamination that does not present a risk for most of the allergic population (95% or 99%, depending on data) according to clinical data available of minimum eliciting doses. Above these reference doses, precautionary labelling warning of potential cross-contamination is required.

VITAL system relies on three key values. First, the "reference amount" that represents the portion size, namely the maximum amount of a food eaten in a typical eating occasion. Second, the "reference dose" which refers to the protein level (total protein in milligrams from an allergenic food) below which only the most sensitive individuals (between 1 and 5) in the allergic population are likely to experience an adverse reaction. Third, the "action levels" that are threshold levels of protein concentrations in food guiding the labelling (action level 1: no precautionary labelling required, action level 2: "may contain" labelling required, and action level 3: "contain" labelling required).

The latter are calculated according to the set reference dose and reference amount, becoming part of action level grids for easy use by food producers.

Starting from this, in order to provide useful tools for food allergen risk-management, good sensitivity is demanded for new analytical methods complying with the action levels prescribed by the VITAL program. Such levels are periodically updated according to new allergological data available from clinical studies, and last values of VITAL program version 3.0, were revised and released in October 2019. As for milk and egg, an equal reference dose of 0.2 $mg_{total\ protein}$ was set, and referred to a portion size of 50 g, deemed reasonable for cookies, thus resulting in an action level 1 of 4 $mg_{total\ protein}$/kg.

Noteworthy, by applying this method following two different routes we obtained two different sensitivities according to the type of allergen contamination occurring in the food matrix. As for incurred cookies the method reached sensitivity down to 4 $mg_{total\ protein}$/kg this one being the minimum level detectable by the method in use and in compliance with the VITAL sensitivities required. This limit might then represent the highest protection level offered to the allergic patient since it refers to a cookie incurred at the beginning of the whole process taking into account the processing effect as well as the extraction efficiency of the containing proteins.

3.4. Precision

Intra-day and inter-day precision of the analytical method (percent coefficient of variation in peak areas at a fixed concentration, CV%) were evaluated to test the method repeatability and reproducibility within the same laboratory. To this purpose, a blank cookie sample fortified with skim milk and whole egg powders at the final level of 100 $\mu g_{allergenic\ ingredient}/g_{matrix}$ was prepared. The intra-day repeatability was calculated within five independent replicates and values lower than 10% were obtained in all cases, with the best repeatability provided by the $\alpha S1$-casein marker FFV and the ovalbumin marker ISQ, due to the high abundance of these proteins in the allergic ingredients. On the contrary, inter-day repeatability was calculated over 3 days by analyzing the same fortified samples.

Obtained values were always lower than 9% for both milk and egg quantifier peptides. The mean values obtained on different days were compared by a one-way ANOVA test at 95% confidence level, resulting in no significant differences for all peptide markers.

3.5. Evaluation of Processing Effect on Method Sensitivity

As known, food processing can deeply affect the structure and stability of a protein as well as its solubility due to several chemical modifications that can occur during thermal treatment. Consequently, the analytical detection can be affected as well when extensively processed foods are investigated for allergen contamination. In order to evaluate the effects of food processing on the detection of each milk and egg peptide marker, specific matrix matched calibration curves were obtained by progressive dilution of incurred cookies extract fortified with milk and egg allergens at high level. As known, incurred material is produced by adding allergic ingredients during dough preparation and before thermal treatment. This condition reproduces what is actually happening during food processing leading to a more reliable estimation of method sensitivity considering the overall effects of processing on protein stability and solubility. As a result, the final recovery and performance of the method could be taken into account. As detailed in Section 2, incurred-cookie calibration curves were produced within a certain concentration range. Following, on the basis of the total protein contents estimated for skim milk powder and whole egg materials used for the preparation of incurred cookies (35.39% and 48.05% of total protein content for milk and egg, respectively) all the peptide reporting units were converted into $\mu g_{total\ protein}/g_{matrix}$ providing an analytical range between 3.5 and 106.2 µg/g for milk proteins and 4.8 and 144.2 µg/g for egg. By linear interpolation of the calculated peak areas we retrieved information on the linearity range and the sensitivity for each precursor/transition acquired. Results are depicted in Table 2. LODs of 1.6 and 3.5 $\mu g_{total\ protein}/g_{matrix}$ were calculated for FFV and TPE quantifier milk peptides while higher LODs were obtained for egg allergen, namely 4.0 and 4.8 $\mu g_{total\ protein}/g_{matrix}$ for ISQ and NIP quantifier egg peptides. By comparing LODs calculated for synthetic peptide-curve calibrations with incurred cookie-curve calibrations, a sensitivity reduction of approximately 94% and 97% was observed both for FFV and ISQ milk and egg peptides due to processing effect, while a slight reduction (of 14% and 38%, respectively) was calculated for TPE (whey proteins) and NIP (yolk proteins) peptides. These results are in accordance with our previous investigation [20] where a reduction of milk and egg detection sensitivity of approximately 93% and 97% were recorded for milk (based on casein marker) and egg allergens (based on white egg marker). The data gathered can be explained by taking into account the labile behavior shown by specific proteins during some processing applied to food [20].

Table 2. Relevant parameters of milk and egg peptides referred to matrix-matched calibration curves built up by using incurred cookies.

Allergenic Ingredient: Protein	Marker	Quantifier Transition	LOD/LOQ ($\mu g_{TOT\ PROT}/g_{MATRIX}$) Incurred Material	R^2
Milk: α-S1-casein	FFV	692.9→991.4	1.6/5.4	0.9969
Milk: β-Lactoglobulin	TPE	623.3→572.5	3.5/11.7	0.9854
Egg: ovalbumin	ISQ	592.1→858.9	4/15.6	0.9903
Egg: vitellogenin	NIP	671.8→557.9	4.8/14.0	0.9896

3.6. Trueness

Trueness of the method was evaluated by performing dedicated experiments on the only reference material available on the market validated for milk detection in cookie matrix. The purchased kit contains two samples at different concentration levels, namely 10 and 50 $\mu g_{allergenic\ ingredient}/g_{matrix}$, that correspond according to the certificate of analysis, to 3.54 and 17.7 $\mu g_{tot\ prot}/g_{matrix}$, respectively.

Trueness evaluation was limited to milk allergen since no reference materials for baked food are available yet for egg allergen. Specifically, incurred reference materials were subjected to the whole sample preparation along with reference allergen-free cookie sample. The latter was used to build-up new matrix-matched calibration curves with synthetic peptides covering the range 1–50 $\mu g_{tot\ prot}/g_{matrix}$, and calculate line equation in the specific cookie matrix provided with the kit. The low and high incurred samples were both analyzed in triplicate (independent samples), and experimental concentration values obtained by curve interpolation were compared with theoretical ones. The percentage ratio between the experimental and theoretical values provided an estimate of the method recovery for milk allergen. Method recovery calculated with YLG peptide was 57 ± 6%, and 56 ± 7% at low and high concentration levels, respectively, whereas the recovery calculated with the peptide FFV was 57 ± 4% and 50 ± 3% at low and high concentration levels, respectively (see Table S3).

3.7. Occurrence of Milk and Egg Contamination in Commercial Samples Declared "Prepared without Adding Milk and Egg"

In the final part of the work, the validated method was applied to samples taken from 10 different lots of commercial cookies and labelled as "prepared without adding of milk and eggs" in order to assess the actual absence of any trace of milk and egg allergens, according to the sensitivity of the method.

Cookies were submitted to sample preparation and analyzed in duplicates with the analytical method herein described and optimized. No quantifiable peaks areas were detected for milk and egg quantifier peptides, therefore, we concluded that no accidental contamination occurred in these samples, at least within the sensitivity limits reported by the developed method. The analytical method in-house validated in the present work, demonstrated to be a sensitive tool for the quantification of egg and milk allergens in cookies at the highest confidence level in compliance with the VITAL doses recommended. This approach, in perspective, can represent a valid alternative to the use of PAL.

4. Conclusions

The method herein described based on QSight triple quadrupole mass analyzer provides an optimized sample preparation protocol and a MRM method for the simultaneous quantification of egg and milk in cookies selected as a model bakery product. Method performance was assessed by using selected milk and egg synthetic peptide markers and a proper factor to convert peptide into protein concentration was proposed. The LOD and LOQ values obtained for both egg and milk allergens calculated in incurred cookies (referred to the protein content) allowed to detect levels of contamination complying with the reference thresholds set for egg and milk and recommended (action level 1) by the VITAL program v 3.0. Additionally, method precision provided good results for both the allergenic ingredients analyzed in this matrix. Processing effects were also assessed confirming previous evidence about the reduced detectability for both allergens, with milk proteins being more susceptible to thermal processing effect. To the best of our knowledge, this is the first time the trueness of the method was calculated by means of MoniQA reference material in this type of food material. The in-house validation performed provided analytical features that complied with the minimum requirements set in the AOAC SMPR 2016.002 for allergen detection in food. Finally, the method was also challenged with real samples from the market to test its realistic potential in detecting accidental cross-contamination in real samples. Commercial cookies labeled as "milk and egg ingredients free" were analyzed by exploiting the developed method and none of them were found incorrectly labeled, within the sensitivity limits achieved with this method. In perspective, the multi-allergen MS based method developed can be employed for allergen control in food supply chains where cross-contamination is likely to occur, hence avoiding the resort to PAL.

Supplementary Materials: The following are available online at http://www.mdpi.com/2304-8158/9/10/1489/s1, Table S1: LC method parameters for milk and egg peptides separation, Table S2: MRM conditions for milk and egg peptides detection, Table S3: Method recovery calculated by using MoniQA reference materials.

Author Contributions: Conceptualization, L.M., I.G.; methodology, E.D.A., R.P., L.M., A.P.G.; validation, E.D.A., R.P.; formal analysis, R.G., E.D.A.; resources, L.M., I.G.; writing—original draft preparation, E.D.A.; writing—review and editing, A.F., L.M., R.P.; supervision, I.G., L.M.; project administration, L.M.; Authorship must be limited to those who have contributed substantially to the work reported. All authors have read and agreed to the published version of the manuscript.

Funding: This research received funding from the Association ALLEGRIA.

Acknowledgments: The authors are grateful to Francesco De Marzo (CNR-ISPA) for his valuable help in instrument purchase and in the preparation of the tender, to Roberto Schena (CNR-ISPA) for his skilled technical support, and to Marinella Cavallo (CNR-ISPA) and Mariella Quarto (CNR-ISPA) for the administrative aid. The authors are also thankful to Alessandro Baldi (PerkinElmer) and Cesare Rossini (Lab Service) for the inputs and the interest in this project.

Conflicts of Interest: The authors declare no conflict of interest.

References

1. Dunlop, J.H.; Keet, C.A. Epidemiology of food allergy. *Immunol. Allergy Clin.* **2018**, *38*, 13–25. [CrossRef]
2. Sicherer, S.H.; Sampson, H.A. Food allergy: A review and update on epidemiology, pathogenesis, diagnosis, prevention, and management. *J. Allergy Clin. Immunol.* **2018**, *141*, 41–58. [CrossRef] [PubMed]
3. Loh, W.; Tang, M.L.K. Debates in Allergy Medicine: Oral immunotherapy shortens the duration of milk and egg allergy–the con argument. *World Allergy Organ. J.* **2018**, *11*, 1–7. [CrossRef] [PubMed]
4. European Regulation (EU) No 1169/Regulation of the European Parliament and of the Council of 25 October 2011 on the provision of food information to consumers. *Off. J. Eur. Union.* **2011**, 18. Available online: https://eur-lex.europa.eu/legal-content/EN/TXT/?uri=CELEX:02011R1169-20180101 (accessed on 1 July 2020).
5. Fierro, V.; Di Girolamo, F.; Marzano, V.; Dahdah, L.; Mennini, M. Food labelling issues in patients with severe food allergies: Solving a hamlet-like doubt. *Curr. Opin. Allergy Clin. Immunol.* **2017**, *17*, 204–211. [CrossRef] [PubMed]
6. Mills, E.N.C.; Adel-Patient, K.; Bernard, H.; De Loose, M.; Gillard, N.; Huet, A.C.; Larré, C.; Nitride, C.; Pilolli, R.; Tranquet, O.; et al. Detection and quantification of allergens in foods and minimum eliciting doses in food-Allergic individuals (ThRAll). *J. Aoac Int.* **2019**, *102*, 1346–1353. [CrossRef] [PubMed]
7. Allen, K.J.; Turner, P.J.; Pawankar, R.; Taylor, S.; Sicherer, S.; Lack, G.; Rosario, N.; Ebisawa, M.; Wong, G.; Mills, E.N.C.; et al. Precautionary labelling of foods for allergen content: Are we ready for a global framework? *World Allergy Organ. J.* **2014**, *7*. [CrossRef]
8. Allen, K.J.; Remington, B.C.; Baumert, J.L.; Crevel, R.W.R.; Houben, G.F.; Brooke-Taylor, S.; Kruizinga, A.G.; Taylor, S.L. Allergen reference doses for precautionary labeling (VITAL 2.0): Clinical implications. *J. Allergy Clin. Immunol.* **2014**, *133*, 156–164. [CrossRef]
9. Taylor, S.L.; Baumert, J.L.; Kruizinga, A.G.; Remington, B.C.; Crevel, R.W.R.; Brooke-Taylor, S.; Allen, K.J.; Houben, G. Establishment of Reference Doses for residues of allergenic foods: Report of the VITAL Expert Panel. *Food Chem. Toxicol.* **2014**, *63*, 9–17. [CrossRef]
10. Cho, C.Y.; Nowatzke, W.; Oliver, K.; Garber, E.A.E. Multiplex detection of food allergens and gluten. *Anal. Bioanal. Chem.* **2015**, *407*, 4195–4206. [CrossRef]
11. Pöpping, B.; Diaz-Amigo, C. The probability of obtaining a correct and representative result in allergen analysis. *Agro Food Ind. Hi Tech* **2014**, *25*, 421–442. [CrossRef]
12. Monaci, L.; De Angelis, E.; Montemurro, N.; Pilolli, R. Comprehensive overview and recent advances in proteomics MS based methods for food allergens analysis. *Trac–Trends Anal. Chem.* **2018**, *106*, 21–36. [CrossRef]
13. Marzano, V.; Tilocca, B.; Fiocchi, A.G.; Vernocchia, P.; Levi Mortera, S.; Urbani, A.; Roncada, P.; Putignani, L. Perusal of food allergens analysis by mass spectrometry-based proteomics. *J. Proteomics.* **2020**, *215*, 103636. [CrossRef]
14. Heick, J.; Fischer, M.; Pöpping, B. First screening method for the simultaneous detection of seven allergens by liquid chromatography mass spectrometry. *J. Chromatogr. A* **2011**, *1218*, 938–943. [CrossRef]

15. Parker, C.H.; Khuda, S.E.; Pereira, M.; Ross, M.M.; Fu, T.J.; Fan, X.; Wu, Y.; Williams, K.M.; DeVries, J.; Pulvermacher, B.; et al. Multi-allergen Quantitation and the Impact of Thermal Treatment in Industry-Processed Baked Goods by ELISA and Liquid Chromatography-Tandem Mass Spectrometry. *J. Agric. Food Chem.* **2015**, *63*, 10669–10680. [CrossRef]
16. Lamberti, C.; Acquadro, E.; Corpillo, D.; Giribaldi, M.; Decastelli, L.; Garino, C.; Arlorio, M.; Ricciardi, C.; Cavallarin, L.; Giuffrida, M.G. Validation of a mass spectrometry-based method for milk traces detection in baked food. *Food Chem.* **2016**, *199*, 119–127. [CrossRef] [PubMed]
17. Planque, M.; Arnould, T.; Dieu, M.; Delahaut, P.; Renard, P.; Gillard, N. Advances in ultra-high performance liquid chromatography coupled to tandem mass spectrometry for sensitive detection of several food allergens in complex and processed foodstuffs. *J. Chromatogr. A* **2016**, *1464*, 115–123. [CrossRef]
18. Planque, M.; Arnould, T.; Dieu, M.; Delahaut, P.; Renard, P.; Gillard, N. Liquid chromatography coupled to tandem mass spectrometry for detecting ten allergens in complex and incurred foodstuffs. *J. Chromatogr. A* **2017**, *1530*, 138–151. [CrossRef]
19. Pilolli, R.; De Angelis, E.; Monaci, L. Streamlining the analytical workflow for multiplex MS/MS allergen detection in processed foods. *Food Chem.* **2017**, *221*, 1747–1753. [CrossRef]
20. Pilolli, R.; De Angelis, E.; Monaci, L. In house validation of a high resolution mass spectrometry Orbitrap-based method for multiple allergen detection in a processed model food. *Anal. Bioanal. Chem.* **2018**, *410*, 5653–5662. [CrossRef] [PubMed]
21. Boo, C.C.; Parker, C.H.; Jackson, L.S. A targeted LC-MS/MS method for the simultaneous detection and quantitation of egg, milk, and peanut allergens in sugar cookies. *J. Aoac Int.* **2018**, *101*, 108–117. [CrossRef] [PubMed]
22. New, S.L.; Schreiber, A.; Stahl-Zeng, J.; Liu, H.F. Simultaneous analysis of multiple allergens in food products by LC-MS/MS. *J. Aoac Int.* **2018**, *101*, 132–145. [CrossRef]
23. Sayers, R.L.; Gethings, L.A.; Lee, V.; Balasundaram, A.; Johnson, P.E.; Marsh, J.A.; Wallace, A.; Brown, H.; Rogers, A.; Langridge, J.I.; et al. Microfluidic Separation Coupled to Mass Spectrometry for Quantification of Peanut Allergens in a Complex Food Matrix. *J. Proteome Res.* **2018**, *17*, 647–655. [CrossRef] [PubMed]
24. Pilolli, R.; Nitride, C.; Gillard, N.; Huet, A.C.; van Poucke, C.; de Loose, M.; Tranquet, O.; Larré, C.; Adel-Patient, K.; Bernard, H.; et al. Critical review on proteotypic peptide marker tracing for six allergenic ingredients in incurred foods by mass spectrometry. *Food Res. Int.* **2020**, *128*, 108747. [CrossRef] [PubMed]
25. Fæste, C.K.; Løvberg, K.E.; Lindvik, H.; Egaas, E. Extractability, stability, and allergenicity of egg white proteins in differently heat-processed foods. *J. Aoac Int.* **2007**, *90*, 427–436. [CrossRef] [PubMed]
26. Korte, R.; Oberleitner, D.; Brockmeyer, J. Determination of food allergens by LC-MS: Impacts of sample preparation, food matrix, and thermal processing on peptide detectability and quantification. *J. Proteom.* **2019**, *196*, 131–140. [CrossRef]

Publisher's Note: MDPI stays neutral with regard to jurisdictional claims in published maps and institutional affiliations.

© 2020 by the authors. Licensee MDPI, Basel, Switzerland. This article is an open access article distributed under the terms and conditions of the Creative Commons Attribution (CC BY) license (http://creativecommons.org/licenses/by/4.0/).

Article

Increasing Coverage of Proteome Identification of the Fruiting Body of *Agaricus bisporus* by Shotgun Proteomics

Tae-Ho Ham [1], Yoonjung Lee [1], Soon-Wook Kwon [2], Myoung-Jun Jang [3], Youn-Jin Park [4] and Joohyun Lee [1,*]

1. Department of Crop Science, Konkuk University, Seoul 05029, Korea; lion78@daum.net (T.-H.H.); yoon10.lee@gmail.com (Y.L.)
2. Department of Crop Plant Bioscience, Pusan National University, Milyang 50463, Korea; swkwon@pusan.ac.kr
3. Department of Plant Resources, Kongju National University, Yesan 32439, Korea; plant119@kongju.ac.kr
4. Kongju National University Legumes Green Manure Resource Center, Yesan 32439, Korea; cocono@naver.com
* Correspondence: edmund@konkuk.ac.kr

Received: 14 April 2020; Accepted: 7 May 2020; Published: 14 May 2020

Abstract: To increase coverage of protein identification of an *Agaricus bisporus* fruiting body, we analyzed the crude protein fraction of the fruiting body by using a shotgun proteomics approach where 7 MudPIT (Multi-Protein identification Technology) runs were conducted and the MS/MS spectra from the 7 MudPIT runs were merged. Overall, 3093 non-redundant proteins were identified to support the expression of those genes annotated in the genome database of *Agaricus bisporus*. The physicochemical properties of the identified proteins, i.e., wide pI value range and molecular mass range, were indicative of unbiased protein identification. The relative quantification of the identified proteins revealed that K5XI50 (Aldedh domain-containing protein) and K5XEW1 (Ubiquitin-like domain-containing protein) were highly abundant in the fruiting body. Based on the information in the Uniprot (Universal Protein Resource) database for *A. bisporus*, only approximately 53% of the 3093 identified proteins have been functionally described and approximately 47% of the proteins remain uncharacterized. Gene Ontology analysis revealed that the majority of proteins were annotated with a biological process, and proteins associated with coiled-coil (12.8%) and nucleotide binding (8.21%) categories were dominant. The Kyoto Encyclopedia of Genes and Genome analysis revealed that proteins involved in biosynthesis of secondary metabolites and tyrosine metabolism were enriched in a fruiting body of *Agaricus bisporus*, suggesting that the proteins are associated with antioxidant metabolites.

Keywords: *Agaricus bisporus*; shotgun proteomics; gene ontology

1. Introduction

The button mushroom *Agaricus bisporus*, a species of macrofungi of the phylum Basidiomycota, is a typical edible fungus, together with the oyster mushroom. Since the start of *A. bisporus* cultivation in the 1650s, several methods of cultivation have been developed [1]. Currently, this mushroom is widely cultivated and consumed globally, particularly in the Netherlands and USA. In Korea, the cultivation of *A. bisporus* was introduced in the 1950s. At present, this mushroom accounts for approximately 7% of the mushroom cultivation industry in Korea, with an annual output ranged 6678 to 13,052 tons in last 10 years [2].

Sustainable development of the mushroom industry in Korea requires breeding new cultivars with superior traits that meet the demands of producers and consumers. Genetic studies of the traits that are fundamental for breeding programs, however, reveal a suppressed recombination frequency of co-segregating markers, resulting in limited and unsaturated genetic linkage maps [3]. Recent genomic methods constitute a possible alternative approach for gaining useful genetic information for the *A. bisporus* breeding program. The haploid genome sequence of *A. bisporus* is available; the genome is comprised by 13 chromosomes and is 30.4 Mb in size [4]. This information has enabled various studies of the genome structure and gene expression of *A. bisporus*.

Proteomics is a powerful tool used for the detection and identification of gene expression and constitution, respectively [5]. With the advances in mass spectrometry technology, proteomics is becoming increasingly relevant for the study of fungal biology, especially the physiology and development of filamentous fungi [5]. As research into *A. bisporus* has started to attract attention, proteome studies have also advanced from the determinations of amino acid composition [6] to two-dimensional polyacrylamide gel electrophoresis (2-D PAGE) [7,8]. Since its original publication, the O'Farrell method for two-dimensional electrophoresis of proteins has been widely used in proteomics [9]. The advantages of this method are that it is easy to apply, and the data are obtained within a short timeframe. However, 2-D PAGE is not suitable for the identification of low-abundance proteins; high-molecular weight or small-molecular weight proteins; hydrophobic proteins; or proteins with a basic pI value, which is the pH at which the net charge of the protein is zero [10]. To overcome the limitations of 2D-PAGE, relevant shotgun proteomics methods recently emerged [11]. Shotgun proteomics, also known as MudPIT, allows for large-scale identification of proteins, which is an area where 2D-PAGE does not perform well [12]. At the same time, shotgun proteomics takes a lot less time than 2D-PAGE analysis [13].

Previous studies of the button mushroom have mainly focused on its cultivation or breeding [14]. Shotgun proteomics in crops has only been used to analyze a few model fungi because this method relies on an available complete genome. In the current study, to provide basic -omics information for *A. bisporus*, we inferred the proteome of a fruiting body of the species *A. bisporus* using shotgun proteomics approaches though merging 7 MudPIT runs.

2. Materials and Methods

2.1. Cultivation of A. bisporus

The cultivar of Sae-han (strain ASI1350), which was obtained from Gyeongsangbuk-do Agricultural Research and Extension Services in Korea (http://www.gba.go.kr/) was prepared as a grain spawn. Mushrooms were grown in a plastic bag filled with 2 kg of the button mushroom compost. The compost was prepared from wheat straw (65% total dry weight), poultry manure (28%), gypsum (4%), limestone (2%), and urea (1%) at pH 8.1 and with 68% moisture. The grain spawn was used to inoculate the compost at 1% concentration. This was followed by a 20 d incubation at 22–23 °C and 60–70% relative humidity in darkness. After the spawn-run, the compost was colonized by 80–90% mycelium of *A. bisporus* and then covered with a 3–4 cm deep layer of the casing soil. The temperature in the cultivation room was adjusted to 17 °C, and the relative humidity was maintained at approximately 90%. The room was ventilated to induce fruiting body formation when the mycelium reached the surface of the casing layer under 14 h light and 10 h dark conditions.

2.2. Protein Extraction

For 7 MudPIT runs, the harvested 7 fruiting bodies (cap) of *A. bisporus* were stored at −80 °C, powdered in liquid nitrogen, and placed in a 1.5 mL Eppendorf tube. Extraction buffer (8 M urea, 5 mM dithiothreitol, 1% LDS, and 100 mM Tris, pH 8.5) was then added, and the powdered samples were homogenized in the buffer. The homogenized samples were centrifuged at 14,000× g for 15 min at 4 °C, and the supernatant was transferred to a new tube. The supernatant was then filtered through membrane filters (0.45 μm size), and the protein was precipitated overnight in the presence of 20% (v/v) trichloroacetic acid. The pellet was washed several times with cold acetone to remove pigments. The protein was then resolubilized in a resolubilization buffer (8 M urea and Tris-HCl, pH 8.5). The concentration of the protein was determined by using 2D-protein Quant kit (GE Healthcare, Piscataway, NJ, USA) as described elsewhere [15].

2.3. Protein Digestion

For the experiment, 300 μg of the sample protein was reduced using Tris (2-carboxyethyl) phosphine hydrochloride (TCEP), by adjusting the concentration of TCEP to 5 mM and incubating for 30 min at room temperature. The reduced sample was carbamidomethylated in the presence of 10 mM iodoacetamide during incubation for 30 min at room temperature in the dark. The protein sample was then diluted with 100 mM Tris-HCl to reduce the urea concentration to 2 M, and $CaCl_2$ was added to a final concentration of 2 mM. Then, 5 μg trypsin and trypsin buffers were added. The sample was incubated overnight at 37 °C to allow for protein digestion. The digested proteins were desalted by passing through a SPEC PLUS PT C18 column (Agilent Technologies, Santa Clara, CA, USA), and the solvent was evaporated by using a speed-Vac.

2.4. Liquid Chromatography–Tandem Mass Spectrometry (LC-MS/MS) Analysis

Nano LC connected to Finnigan LTQ mass spectrometer (Thermo Scientific, Waltham, MA) was used. Biphasic columns were prepared in-house for analysis; the columns contained 365 μm OD (outer diameter) × 100 μm ID (inside diameter) fused-silica capillaries (Polymicro Technologies, Phoenix, AZ, USA). The capillaries were packed using a pressure cell with helium and 9 cm of C18-AQ 5 μm reverse phase (PP), followed by 3 cm of 5 μm strong cation exchange Luna resin. The desalted sample was loaded onto the in-house column in 20 μL of a mixture of 5% acetonitrile and 0.1% formic acid.

Reversed-phase chromatography was performed using a binary buffer system of 0.1% formic acid (buffer A) and acetonitrile in 0.1% formic acid (buffer B). Nano LC was performed by using a linear gradient of 3–50% of buffer B at a flow rate of 0.200 μL/min. The peptides were eluted in the course of an 11-step program of increasing concentration of salt solution. The eluent was ionized by electrospraying from the column directly into the MS/MS system. A parent-ion scan was performed in the range of 400–1600 m/z (mass-to-charge ratio). The top five most intense parent ions were chosen, and an MS/MS-ion scan was performed by collision-induced dissociation. The run time for LC-MS/MS was 120 min for each step. The total run time for the 11 steps was approximately 22 h. A total of 7 MudPIT runs were conducted.

2.5. Proteomic Data Analysis

The MS/MS spectral files for 7 MudPIT runs were merged in the single file. For each identified protein, protein ID, spectra count, pI value, and molecular mass were determined by using the Proteome Discoverer software (version 1.3) (Thermo Fisher Scientific, Waltham, MA, USA) with the merged single MS/MS spectra file. The fragmentation spectra data for *A. bisporus* were searched using Uniprot (http://www.uniprot.org/). For database searching, carbamidomethylation of cysteine was set as a fixed modification, and oxidation of methionine was set as a variable modification. To verify these modifications, protein identifications were filtered at 1% false discovery rate.

2.6. Determination of the Relative Protein Abundances

The output of the Proteome Discoverer analysis was exported to Microsoft Excel to calculate the normalized spectral counts (NSpC). The NSpC for each protein k is given by the equation

$$(\text{NSpC})_k = \frac{\left(\frac{\text{SpC}}{\text{L}}\right)_k}{\sum_{i=1}^{n}\left(\frac{\text{SpC}}{\text{L}}\right)_i} \quad (1)$$

where the total number of MS/MS spectra matching peptides from the protein k (SpC) is divided by the protein's length (L) and then divided by the sum of all SpC/L values for the protein in the experiment.

2.7. Bioinformatics Analysis

Gene Ontology (GO) annotations for *A. bisporus* were from Uniprot (http://www.uniprot.org/). GO analysis and Kyoto Encyclopedia of Genes and Genome (KEGG) pathway analysis was performed using the DAVID (Database for Annotation, Visualization and Integrated Discovery, version 6.8) (https://david.ncifcrf.gov/) and the *A. bisporus* Uniprot database was used as a reference. Protein GO classification was performed by using PANTHER (Protein ANalysis THrough Evolutionary Relationships, version 14.0, Los Angeles, CA, USA) (http://www.pantherdb.org/) classification system and CateGOrizer (http://www.animalgenome.org/tools/catego/). The enriched function-related proteins were grouped. Transmembrane domains were predicted using TMHMM (version 2.0, Lyngby, Denmark) (http://www.cbs.dtu.dk/services/TMHMM/) with protein sequence data from Uniprot.

3. Results and Discussion

3.1. Identified Proteins of the A. bisporus Fruiting Body

Overall, the shotgun proteomic analysis of the fruiting body of the species *A. bisporus* identified 3093 proteins (Supplementary Table S1). Compared with the 2D-PAGE technology that can generally resolve approximately 1000 protein spots in a single gel, the shotgun proteomic analysis conducted in the current study had a much higher resolving power, identifying ~1000 proteins for a single MudPIT run. A phenomenon of analytical incompleteness in MudPIT analysis was reported previously [16], in which any single analytical run may only identify a fraction of the relevant peptides in a highly complex mixture of peptides. Therefore, for deep proteome identification, we conducted 7 MudPIT runs with the fruiting bodies of the Sae-han cultivar, and the MS/MS spectra were merged to a single file. From the protein identification with the single merged MS/MS file, we identified 3093 non-redundant proteins of the fruiting body of *A. bisporus*. After 5 MudPIT runs were merged, the number of newly identified proteins was less than 100 as 6, 7 MudPIT runs were merged. Thus, to increase the coverage of protein identification more than our results, other strategies are required such as protein sample fractionation or solubilizing hydrophobic proteins. In addition to increasing the protein identification coverage, the confidence level of the protein identification also increased. At least one unique peptide was assigned to the identified proteins. The spectral counts that represent the number of MS/MS spectra assigned to a peptide from certain proteins were found to have increased the identified proteins. The minimum number of spectral counts for the identified protein was 81 for K5Y688 (CS domain-containing protein) and the maximum was 16,914 for K5XHR7 (uncharacterized protein). Around 80% of the identified proteins showed more than 1000 spectral counts. (Supplementary Table S1). Spectral counting has become a commonly used approach for measuring relative abundance of proteins in label-free shotgun proteomics [17]. The increased number of the spectral count for the identified pterions improved acquiring relative abundance of proteins.

The distributions of the physiochemical properties of the identified proteins, pI values, and molecular masses were analyzed. The pI value and molecular mass are the main and limiting factors that affect protein resolution in 2D-PAGE because of the gel properties. The distributions of the pI values and molecular masses of the identified proteins were compared with those of proteins predicted from the whole *A. bisporus* genome sequence (Figure 1A,B).

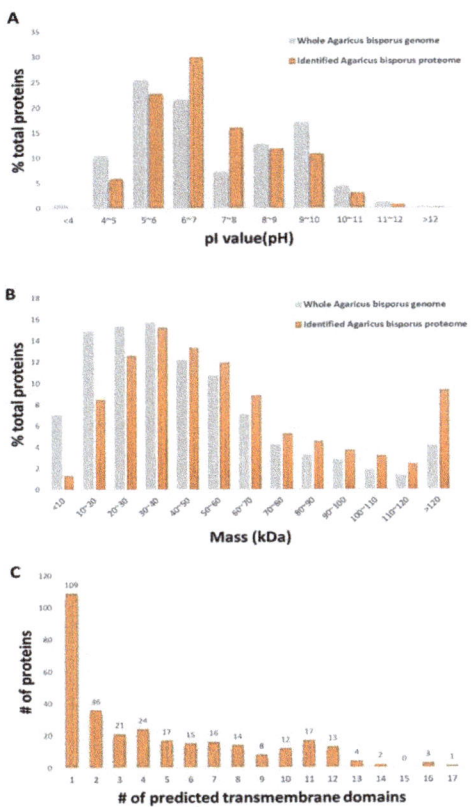

Figure 1. Bioinformatic analysis of the identified proteins. (**A**) Distributions of isoelectric point (pI). (**B**) Distributions of molecular mass. (**C**) Number of predicted transmembrane domains.

K5XUF0 (uncharacterized protein) had the highest pI value at 12.19. K5VUP2 (uncharacterized protein) had the lowest pI value at 4.01. Based on the distribution of the pI values of the identified proteins, approximately 42% of proteins had a pI higher than 7.0, which, in general, would limit their resolution on 2D-PAGE. No proteins with a pI value below 4 were detected, even though such proteins are putatively encoded by the *A. bisporus* genome. That is probably because of the low proportion of acidic proteins encoded by the fungal genome. The molecular mass of the identified proteins ranged from 5.6 kDa (K5XCV4; uncharacterized protein) to 547.6 kDa (K5VXU2; Midasin). Based on the molecular mass data, the mass distribution of the identified proteome was similar to that of the predicted whole proteome. However, the proportion of identified proteins with a molecular mass lower than 30 kDa was smaller than that predicted from the whole genome, and the proportion of proteins with a molecular mass higher than 120 kDa was higher than that predicted from the whole genome. The distribution of the identified proteins showing a wide range of molecular weight and isoelectric points suggested that there was little bias or information loss with the methods used in deep proteome identification of the fruiting body of the species *A. bisporus*. Protein having transmembrane domains predicted by Transmembrane Helices Hidden Markov Models (TMHMM) were explored (Figure 1C, Supplementary Table S2). Based on our protein extraction and solubilization methods, the extraction method was specific for soluble proteins. Even though neither cellular organelles nor cellular membrane were purified in this experiment, 312 proteins were predicted to have transmembrane domain. Interestingly, the functions of all of the 312 proteins were unknown. This result represents insufficient database accumulation for the gene functions encoded in the genome of *A. bisporus*. Based on the fact that the prediction programs used is not perfect, some of the predicted proteins may not be membrane bounded proteins; however, this result suggests that some of the membrane bound proteins are possibly identified by the MudPIT analysis applied in our experiment without additional membrane fractionation procedure.

3.2. Highly Abundant Proteins in the Fruiting Body of A. bisporus

We next used the spectral counts method [18] to determine the relative abundance of the identified proteins. The method considers the spectral counts in the MS/MS spectra data, normalized as NSpC, to determine the relative protein abundance as a proportion of the sum of the relative abundances of all the identified proteins. The abundance of the identified proteins was compared with NSpC. Even though the statistical analyses were not applied in this comparison, the high score of NSpC represents its high abundance in the cell of the fruiting body. Unlike the plant leaf proteome, in which RubisCO and photosynthesis-associated proteins are highly dominant, no highly dominant proteins were apparent in the fruiting body of the species *A. bisporus*. The abundance of the top ten highly abundant proteins accounted for 3.0% of all proteins (Table 1). Among the 3093 identified proteins, the most abundant protein was K5XI50 (Aldedh domain-containing protein), which is the conserved domain of the Aldehyde Dehydrogenase (ALDH) superfamily, which is known to be involved in metabolism and abiotic/biotic stress responses [19]. K5XEW1 (Ubiquitin-like domain-containing protein) was another highly abundant protein in the fruiting body. This protein is found across the Eukarya and is involved in the regulation of signal transduction and enzymatic activity [20]. Ribosomal proteins such as ribosomal_S10 domain-containing protein and SBDS domain-containing protein were also abundant in the fruiting body. Interestingly, transcription elongation factor TEF EF1B localized in the nucleus was highly present.

3.3. Functional Classification and GO Analysis

Based on the information for *A. bisporus* in the Uniprot database, we were able to determine a function for 53% of the 3093 identified proteins whereas for 47% we were unable to determine their functions. Among the top ten highly abundant proteins, only five were functionally described. This indicates that even though the completed genome sequence information is available, follow-up studies, such as functional annotation and gene expression, are still required. The detection of uncharacterized proteins in the fruiting body of the species *A. bisporus* in the current study supports their possible role and presence in the fruiting body. GO information was available for 2543 of the identified proteins. We used GO analysis to categorize the proteins into 89 groups according to their GO functions. The highest proportion of proteins represented the category biological process (13.3%), followed by molecular function (11.9%), catalytic activity (8.2%), metabolism (8.1%), and cellular component (5.4%) (Figure 2).

We then conducted GO enrichment analysis to compare the proportion of the GO terms for the predicted proteins encoded by the genome and the proteins identified in the fruiting body. In the proteins identified in the fruiting body, 20 GO terms for biological processes, 15 GO terms for cellular components, and 8 GO terms for molecular functions were recovered (Figure 3).

Table 1. Top 10 highly abundant proteins in the fruiting body of *Agaricus bisporus*.

Accession	Description	Score	# Proteins	# Peptides	# PSMs	# AAs	MW [kDa]	calc. pI	NSpC
K5XI50	Aldedh domain-containing protein	253.09	2	14	10,609	107	11.60162	5.046387	0.005653
K5XEW1	Ubiquitin-like domain-containing protein	323.58	3	59	10,190	159	18.03673	9.97904	0.003654
K5Y3E9	Uncharacterized protein	232.87	1	45	8737	161	18.49427	4.665527	0.003094
K5VUM2	Ribosomal_S10 domain-containing protein	157.35	1	55	6348	124	13.7715	9.671387	0.002919
K5XEM2	Uncharacterized protein	203.94	1	26	5533	112	11.29655	4.335449	0.002817
K5X7C3	Uncharacterized protein	244.78	1	52	13,010	277	30.24033	5.160645	0.002678
K5X9Z8	SBDS domain-containing protein	212.38	1	29	5666	125	13.97901	6.303223	0.002585
K5Y5Z1	Uncharacterized protein	214.95	1	59	6529	147	16.15384	10.18408	0.002532
K5WMW0	Transcription elongation factor TEF EF1B	282.18	1	48	9256	212	23.10943	4.703613	0.002489
K5XAS4	Uncharacterized protein	313.32	1	76	8200	190	21.39238	9.495605	0.002461

Score: the sum of the ion scores of all peptides that were identified; coverage: the percentage of the protein sequence covered by identified peptide; # proteins: the number of identified proteins in a protein group; # unique peptides: the number of peptide sequences that are unique to a protein group; # peptides: the total number of distinct peptide sequences identified in the protein group; # PSMs: the total number of identified peptide spectra matched for the protein; # AAs: the total number of amino acid in the protein; MW [kDa]: molecular masses of protein; pI: isoelectric points; NSpC: normalized spectral counts.

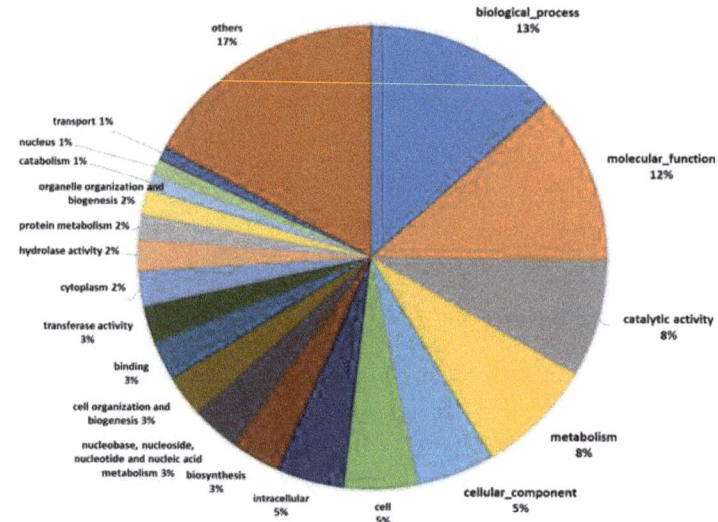

Figure 2. Gene Ontology (GO) functional categorization.

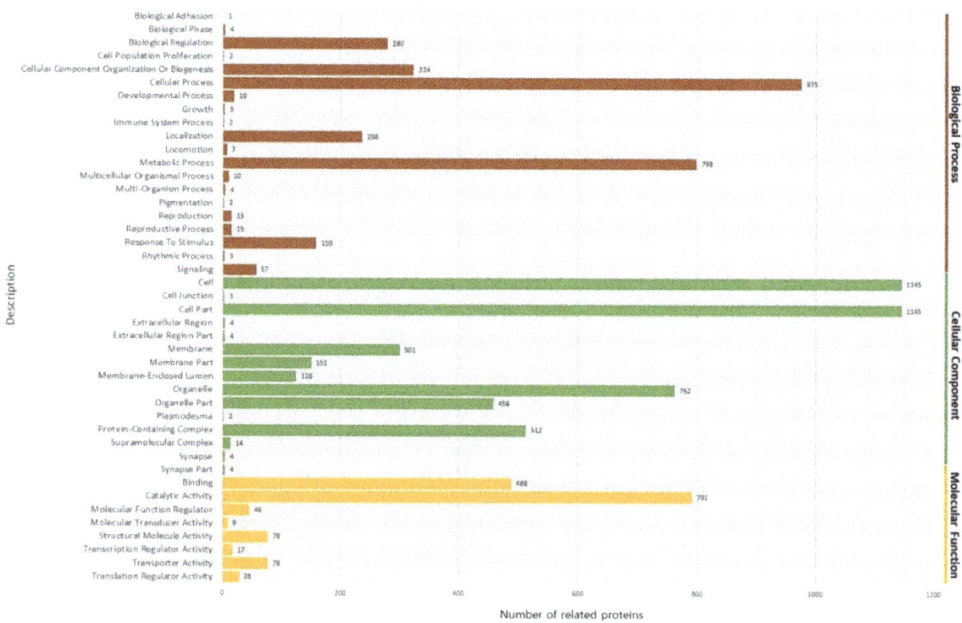

Figure 3. GO analysis describing the three main categories, biological process, cellular component, molecular function.

Among the GO terms for biological processes, the proportion of proteins associated with a cellular process and metabolic process was high. Among GO terms for the cellular components, the proportion of proteins associated with cell compartment and the cell was high. Among the proteins associated with molecular functions, the proportion of proteins associated with catalytic activity and binding was high. When functionally related proteins were grouped, 70 enriched groups were apparent (Supplementary Table S3). The proteins grouped in the coiled-coil (12.8%) and nucleotide binding

(8.21%) categories were dominant. The coiled-coil proteins are thought to facilitate the expansion of the centrosome to aid cell division [21]. The enriched proteins involved in nucleotide binding and coiled-coil proteins may represent the cell division status of the fruiting body of the species *A. bisporus*. Further, the enrichment of proteins related to metabolic pathways and secondary metabolites is possibly associated with the variety of nutrients present in the fruiting body of the species *A. bisporus* [22]. To test how well our dataset covers known cellular pathways, the identified fruiting body proteome was overlaid onto the pathway database of the KEGG [23] (Table 2).

Table 2. KEGG Pathway analysis of identified *Agaricus bisporus* proteins.

KEGG Pathway Term	Number of Related Proteins	Percentage (%)	p-Value
2-Oxocarboxylic acid metabolism	27	1.85	1.70×10^{-3}
Alanine, aspartate and glutamate metabolism	25	1.71	4.40×10^{-4}
Arginine biosynthesis	14	0.96	1.70×10^{-2}
Biosynthesis of amino acids	81	5.54	3.40×10^{-8}
Biosynthesis of antibiotics	148	10.12	1.80×10^{-5}
Biosynthesis of secondary metabolites	186	12.71	1.50×10^{-3}
Carbon metabolism	79	5.40	4.80×10^{-6}
Citrate cycle (TCA cycle)	24	1.64	5.50×10^{-3}
Endocytosis	49	3.35	2.60×10^{-2}
Glycolysis/Gluconeogenesis	36	2.46	2.00×10^{-2}
Metabolic pathways	424	28.98	2.80×10^{-2}
Methane metabolism	19	1.30	3.30×10^{-2}
Propanoate metabolism	10	0.68	3.40×10^{-2}
Proteasome	35	2.39	4.10×10^{-6}
Ribosome	67	4.58	4.70×10^{-3}
RNA transport	69	4.72	1.10×10^{-3}
Spliceosome	61	4.17	2.20×10^{-3}
Tyrosine metabolism	18	1.23	4.60×10^{-2}

Percentage: the proportion of related proteins in relation to total number of proteins; *p*-value: probability of obtaining results.

KEGG pathway enrichment analysis was performed by using DAVID. There were 23 pathways obtained and 18 of them were statistically significant ($p < 0.05$). Among the 3093 proteins, only 23% (731 proteins) were assigned to a KEGG pathway. Of these, the category of metabolic pathway was highly enriched. The highly enriched metabolic pathway in the fruiting body of the species *Auricularia heimuer* has been previously reported in a gene expression study [24]. A total of 186 proteins were assigned to the category of secondary metabolites, and three of them were assigned to terpenoid backbone biosynthesis. Terpenoids are derived from the five-carbon isoprene units and consist of multicyclic structures that differ from one another on the basis of carbon skeleton and different functional groups [25]. Terpenoids isolated from mushrooms have been reported that are associated with various pharmacological activities such as anticancer, anticholinesterase, antiviral, antibacterial, anti-inflammatory, and antioxidase activities [26]. A total of 18 proteins were assigned to the category of tyrosine metabolism. Tyrosine, as a building block, is a precursor for phenolic compounds [27] which are effective antioxidants. Edible mushrooms as a food are sources of protein, fiber, vitamins, minerals, and useful bioactive metabolites with a broad spectrum of pharmacological benefits such as antioxidant activity [28]. Button mushrooms as a food source for an antioxidant can be used to help the organism to reduce oxidative damage [29]. The detected proteins associated in the pathway of antioxidant in the fruiting body of the species *A. bisporus* are possible candidate genes for developing high quality button mushroom cultivars as an antioxidant-containing food.

4. Conclusions

Using a shotgun proteomics approach where merging 7 MudPIT runs increased coverage of protein identification and the confidence level of the identified protein, we identified 3093 proteins in the fruiting body of the species *A. bisporus*. The physiochemical properties of the identified proteins (the pI values and molecular masses) indicated unbiased protein identification. Because of the lack of gene or protein expression data for *A. bisporus*, only approximately 53% of the identified proteins were functionally described and approximately 47% of the proteins are still uncharacterized. The current study is the first report presenting a list of over 3000 proteins in the fruiting body of the species *A. bisporus*. The shotgun proteomics approach and the complete genomic database allowed us to identify more than 2000 non-redundant proteins. Unfortunately, the function of approximately 47% of the identified proteins has not yet been described because genomic and annotation analyses of *A. bisporus* are limited, unlike in other model organisms. However, proteomic detection of these proteins may act as supporting evidence for their existence. The shotgun proteomic approach employed in the current study could be used in further studies comparing the proteome of the fungus *A. bisporus* grown in certain environments or at certain developmental stages.

Supplementary Materials: The following are available online at http://www.mdpi.com/2304-8158/9/5/632/s1, Table S1: The list of total identified proteins of *Agaricus bisporus*. Table S2. The predicted transmembrane domains of the identified *Agaricus bisporus* proteins. Table S3. Table of functionally grouped proteins in *Agaricus bisporus*.

Author Contributions: Methodology, T.-H.H.; software, Y.L.; resources, Y.L.; writing—original draft preparation, T.-H.H., J.L.; writing—review and editing, Y.-J.P., S.-W.K.; visualization, Y.L.; supervision, J.-L.P.; Project Administration, J.L.-P.; funding acquisition, M.-J.J. All authors have read and agreed to the published version of the manuscript.

Funding: This subject is supported by Korea Ministry of Environment as Development of technology to improve the efficiency in transforming food waste into useful resources (2018000710002).

Acknowledgments: We would like to express our gratitude to the Ministry of Environment, Korea for providing the fund to research. Also, thanks to Yebin Kwon and Jeehye Kim for their assistance.

Conflicts of Interest: The authors declare no conflict of interest.

References

1. Oei, P. *Mushroom Cultivation: Appropriate Technology for Mushroom Growers*; Backhuys Publishers: Kerkwerve, The Netherlands, 2003.
2. Yoo, Y.B.; Oh, M.J.; Oh, Y.L.; Shin, P.G.; Jang, K.Y.; Kong, W.-S. Development trend of the mushroom industry. *J. Mushrooms* **2016**, *14*, 13. [CrossRef]
3. Gao, W.; Weijn, A.; Baars, J.J.; Mes, J.J.; Visser, R.G.; Sonnenberg, A.S. Quantitative trait locus mapping for bruising sensitivity and cap color of *agaricus bisporus* (button mushrooms). *Fungal Genet. Biol.* **2015**, *77*, 69–81. [CrossRef] [PubMed]
4. Morin, E.; Kohler, A.; Baker, A.R.; Foulongne-Oriol, M.; Lombard, V.; Nagy, L.G.; Ohm, R.A.; Patyshakuliyeva, A.; Brun, A.; Aerts, A.L.; et al. Genome sequence of the button mushroom *agaricus bisporus* reveals mechanisms governing adaptation to a humic-rich ecological niche. *Proc. Natl. Acad. Sci. USA* **2012**, *109*, 17501–17506. [CrossRef] [PubMed]
5. Kim, Y.; Nandakumar, M.P.; Marten, M.R. Proteomics of filamentous fungi. *Trends Biotechnol.* **2007**, *25*, 395–400. [CrossRef]
6. Braaksma, A.; Schaap, D. Protein analysis of the common mushroom *agaricus bisporus*. *Postharvest Biol. Technol.* **1996**, *7*, 119–127. [CrossRef]
7. Bae, H.-S.; Kim, D.-H.; Choi, U.-K. Proteomic characteristics of calcium enriched king oyster mushroom (*pleurotus eryngii*). *Korean J. Food Sci. Technol.* **2011**, *43*, 12–16. [CrossRef]
8. Horie, K.; Rakwal, R.; Hirano, M.; Shibato, J.; Nam, H.W.; Kim, Y.S.; Kouzuma, Y.; Agrawal, G.K.; Masuo, Y.; Yonekura, M. Proteomics of two cultivated mushrooms *sparassis crispa* and *hericium erinaceum* provides insight into their numerous functional protein components and diversity. *J. Proteome Res.* **2008**, *7*, 1819–1835. [CrossRef]

9. O'Farrell, P.H. High resolution two-dimensional electrophoresis of proteins. *J. Biol. Chem.* **1975**, *250*, 4007–4021.
10. Harry, J.L.; Wilkins, M.R.; Herbert, B.R.; Packer, N.H.; Gooley, A.A.; Williams, K.L. Proteomics: Capacity versus utility. *Electrophoresis* **2000**, *21*, 1071–1081. [CrossRef]
11. Haynes, P.A.; Roberts, T.H. Subcellular shotgun proteomics in plants: Looking beyond the usual suspects. *Proteomics* **2007**, *7*, 2963–2975. [CrossRef]
12. Agrawal, G.K.; Jwa, N.S.; Rakwal, R. Rice proteomics: Ending phase i and the beginning of phase ii. *Proteomics* **2009**, *9*, 935–963. [CrossRef] [PubMed]
13. Lee, J.; Cooper, B. Alternative workflows for plant proteomic analysis. *Mol. Biosyst.* **2006**, *2*, 621–626. [CrossRef] [PubMed]
14. Chandran Sathesh-Prabu, Y.-K.L. Mutation breeding of mushroom by radiation. *J. Radiat. Ind.* **2011**, *5*, 11.
15. Lee, J.; Garrett, W.M.; Cooper, B. Shotgun proteomic analysis of arabidopsis thaliana leaves. *J. Sep. Sci.* **2007**, *30*, 2225–2230. [CrossRef]
16. Wilkins, M.R.; Appel, R.D.; Van Eyk, J.E.; Chung, M.C.M.; Gorg, A.; Hecker, M.; Huber, L.A.; Langen, H.; Link, A.J.; Paik, Y.K.; et al. Guidelines for the next 10 years of proteomics. *Proteomics* **2006**, *6*, 4–8. [CrossRef]
17. Choi, H.; Fermin, D.; Nesvizhskii, A.I. Significance analysis of spectral count data in label-free shotgun proteomics. *Mol. Cell. Proteom.* **2008**, *7*, 2373–2385. [CrossRef]
18. Liu, H.; Sadygov, R.G.; Yates, J.R. A model for random sampling and estimation of relative protein abundance in shotgun proteomics. *Anal. Chem.* **2004**, *76*, 4193–4201. [CrossRef]
19. Asiimwe, T.; Krause, K.; Schlunk, I.; Kothe, E. Modulation of ethanol stress tolerance by aldehyde dehydrogenase in the mycorrhizal fungus *tricholoma vaccinum*. *Mycorrhiza* **2012**, *22*, 471–484. [CrossRef]
20. Grabbe, C.; Dikic, I. Functional roles of ubiquitin-like domain (uld) and ubiquitin-binding domain (ubd) containing proteins. *Chem. Rev.* **2009**, *109*, 1481–1494. [CrossRef]
21. Kuhn, M.; Hyman, A.A.; Beyer, A. Coiled-coil proteins facilitated the functional expansion of the centrosome. *PLoS Comput. Biol.* **2014**, *10*, e1003657. [CrossRef]
22. Muszynska, B.; Kala, K.; Rojowski, J.; Grzywacz, A.; Opoka, W. Composition and biological properties of *agaricus bisporus* fruiting bodies—A review. *Pol. J. Food Nutr. Sci.* **2017**, *67*, 173–181. [CrossRef]
23. Huang, D.W.; Sherman, B.T.; Lempicki, R.A. Systematic and integrative analysis of large gene lists using david bioinformatics resources. *Nat. Protoc.* **2009**, *4*, 44–57. [CrossRef] [PubMed]
24. Yuan, Y.; Wu, F.; Si, J.; Zhao, Y.-F.; Dai, Y.-C. Whole genome sequence of *Auricularia heimuer* (basidiomycota, fungi), the third most important cultivated mushroom worldwide. *Genomics* **2019**, *111*, 50–58. [CrossRef] [PubMed]
25. Jan, S.; Abbas, N. Chapter 4—Chemistry of himalayan phytochemicals. In *Himalayan Phytochemicals*; Jan, S., Abbas, N., Eds.; Elsevier: Amsterdam, The Netherlands, 2018; pp. 121–166.
26. Duru, M.E.; Cayan, G.T. Biologically active terpenoids from mushroom origin: A review. *Rec. Nat. Prod.* **2015**, *9*, 456–483.
27. Feduraev, P.; Skrypnik, L.; Riabova, A.; Pungin, A.; Tokupova, E.; Maslennikov, P.; Chupakhina, G. Phenylalanine and tyrosine as exogenous precursors of wheat (*triticum aestivum* L.) secondary metabolism through PAL-associated pathways. *Plants* **2020**, *9*, 476. [CrossRef] [PubMed]
28. Lindequist, U.; Niedermeyer, T.H.; Julich, W.D. The pharmacological potential of mushrooms. *Evid. Based Complement. Alter. Med.* **2005**, *2*, 285–299. [CrossRef] [PubMed]
29. Singla, R.; Ganguli, A.; Ghosh, M. Antioxidant activities and polyphenolic properties of raw and osmotically dehydrated dried mushroom (*agaricus bisporous*) snack food. *Int. J. Food Prop.* **2010**, *13*, 1290–1299. [CrossRef]

© 2020 by the authors. Licensee MDPI, Basel, Switzerland. This article is an open access article distributed under the terms and conditions of the Creative Commons Attribution (CC BY) license (http://creativecommons.org/licenses/by/4.0/).

Review

Proteomic Insights into the Biology of the Most Important Foodborne Parasites in Europe

Robert Stryiński [1,*], Elżbieta Łopieńska-Biernat [1] and Mónica Carrera [2,*]

1. Department of Biochemistry, Faculty of Biology and Biotechnology, University of Warmia and Mazury in Olsztyn, 10-719 Olsztyn, Poland; ela.lopienska@uwm.edu.pl
2. Department of Food Technology, Marine Research Institute (IIM), Spanish National Research Council (CSIC), 36-208 Vigo, Spain
* Correspondence: robert.stryinski@uwm.edu.pl (R.S.); mcarrera@iim.csic.es (M.C.)

Received: 18 August 2020; Accepted: 27 September 2020; Published: 3 October 2020

Abstract: Foodborne parasitoses compared with bacterial and viral-caused diseases seem to be neglected, and their unrecognition is a serious issue. Parasitic diseases transmitted by food are currently becoming more common. Constantly changing eating habits, new culinary trends, and easier access to food make foodborne parasites' transmission effortless, and the increase in the diagnosis of foodborne parasitic diseases in noted worldwide. This work presents the applications of numerous proteomic methods into the studies on foodborne parasites and their possible use in targeted diagnostics. Potential directions for the future are also provided.

Keywords: foodborne parasite; food; proteomics; biomarker; liquid chromatography-tandem mass spectrometry (LC-MS/MS)

1. Introduction

Foodborne parasites (FBPs) are becoming recognized as serious pathogens that are considered neglect in relation to bacteria and viruses that can be transmitted by food [1]. The mode of infection is usually by eating the host of the parasite as human food. Many of these organisms are spread through food products like uncooked fish and mollusks; raw meat; raw vegetables or fresh water plants contaminated with human or animal excrement. Most FBPs are related to outdated farming procedures and/or to wild animals [2]. Some food is contaminated by food service employees who do not follow sanitation rules or work in unsanitary facilities. The globalization of food supply, the increase of international trade, the convenience of travel, the increase of highly susceptible people (such as aging, malnutrition, human immunodeficiency virus infection), changes in cooking traditions and lifestyles, and advanced diagnostic tools are some of the reasons for the increase in the incidence of food-borne parasitic diseases worldwide [3–5].

In a general context, it was confirmed that there might be up to 50% more species benefit from parasitic lifestyle than all other feeding strategies [6]. Parasites, in particular protozoa (Protozoa), roundworms (Nematoda), flukes (Trematoda), and tapeworms (Cestoda) are enormously different types of eukaryotes that may cause human infection. Their complex lifecycles; varied transmission routes, including water, soil, food, and contacts between people or between animals and people; as well as prolonged periods between infection and symptoms have resulted in their receiving considerable attention in the last few decades [3,7]. The most examined species, *Homo sapiens*, can serve as host to 342 different helminth species and to 70 more if we count the Protozoa. According to local geographic, ecological and economic conditions, every human population in the world has its own unique suit of parasites [8]. The development of animal husbandry, sanitary conditions and diagnostic methods has undoubtedly reduced or even eliminated certain parasite species in industrialized countries and some developing countries. However, the decrease in the number of cases is not common for all parasitic species, especially for foodborne parasites, and there are countries where the occurrence of these infections in humans is still high [2].

Monitoring of foodborne diseases is a fundamental component of food-safety systems. The European Union has introduced regulations for some FBPs, such *Trichinella* spp., *Taenia* spp. and *Anisakis* spp. [9–11]. There are currently no European Union standards published exclusively for Protozoa in food products. However, after massive cryptosporidiosis outbreaks, the food industry began to pay attention to *Cryptosporidium* spp. [12–14]. Validated methods are essential to ensure robust detection of FBPs. The current guidelines are used to monitor bacteria and their direct application to FBPs is not possible. Other concerns include the large differences in FBPs populations (from protozoa to parasitic worms) and their biological differences (for example, different transmission routes, complex development cycles). The detection procedures of FBPs in food products are also different. There are also differences in the range of foods that FBPs may exist and be delivered to potential human hosts. Because of the complex development cycle of many FBPs and the wide variety of hosts, the food analyzed for the detection of a specific FBP can include uncooked meat, fish and other seafood, and fruits or vegetables. [15]. In these cases, research is needed to identify new, more specific treatment targets.

In recent years, proteomics methods have become more and more popular in the food science community [16–19]. New methods for detecting parasites are still an urgent research matter that can successfully benefit from proteomic methodologies.

Proteomics is defined as "the large-scale functional analysis of gene products or functional genomics, including identification or localization studies of proteins" [20]. Proteomics methods are used for the identification and quantification of the protein composition of cells, subfractions of cells, or the medium or secretome surrounding cells at a certain time, collectively termed "the proteome," but also to describe protein modifications and interactions [17]. Proteomic analysis involves the extraction, purification and fractionation of proteins which are identified using mass spectrometry (MS) [21]. Earlier proteomic studies generally used two-dimensional gel electrophoresis (2-DE) approach separating protein mixtures according to charge (pI) and molecular weight (MW), after which proteins could be identified using MS [22]. Today, bottom-up or "shotgun" proteomic approaches that analyze proteins after proteolytic digestion can be coupled to liquid chromatography tandem mass spectrometry (LC-MS/MS), which allows for the high-throughput quantitation of the proteome [23]. These require a high quality and representative genome sequence to map thousands of MS spectra of peptides back to their proteins for identification and quantitation and allow for the characterization of whole proteomes. These conventional approaches were not possible to use in parasitological studies until now. These days, the full or extensive nuclear genome coverage for many parasites of agricultural, veterinary, and medical importance [24] makes proteomics more interesting for parasitologists.

In order to confirm the identity of novel proteins and their function, the application of proteomics is a key to enable extensive characterizations of them or, more widely, the structures and organelles where those proteins are expressed and which are essential in pathogenicity but which are in many cases lacking in more widely studied model organisms, for instance yeast or *Caenorhabditis elegans*–free-living nematode [25]. To fully understand the molecular mechanisms related to the pathogenic characteristics of FBPs, it is most important to analyze the surface proteins and the proteins in extracellular vesicles that represent the frontier of interaction between the parasite and its host. [26–29]. In the past decade, people have used the latest technological advances in proteomics and bioinformatics to develop proteomics strategies to analyze this complex class of molecules [19]. The identification of parasite-specific proteins can significantly simplify the design of new tools for fast and inexpensive diagnosis, which in turn can help break the spread of parasites. In addition, identifying potential vaccination targets (proteins) appears to be one of the leading ways to control parasitic diseases [30]. Accurate knowledge and description of the mechanism of action of these proteins can be used in the research on antiparasitic drugs, and help in combating FBPs through detection and/or neutralization [31]. At the same time, scientists are making great efforts to clarify the sensitization mechanism of various allergenic proteins from food sources, where allergic reactions to food are more often caused by FBP allergens contaminating food products [3,32,33].

Detection of specific parasites in humans and the etiological cause of the disease, like food products, required scientists to employ two characteristic approaches using proteomic methods, i.e., discovery and targeted workflows [17,18,34]. Discovery proteomics is applied to identify and characterize the proteins of FBPs (e.g., global proteomes, cellular, subcellular, or excretory-secretory proteomes) usually employing previously mentioned, bottom-up methodology. Targeted proteomics are based on the monitoring of the protein biomarkers (single or multiple peptides) in analyzed samples, e.g., food products. In targeted proteomics, selected/multiple/parallel-reaction monitoring (SRM/MRM/PRM) is preferably used [35] (Figure 1).

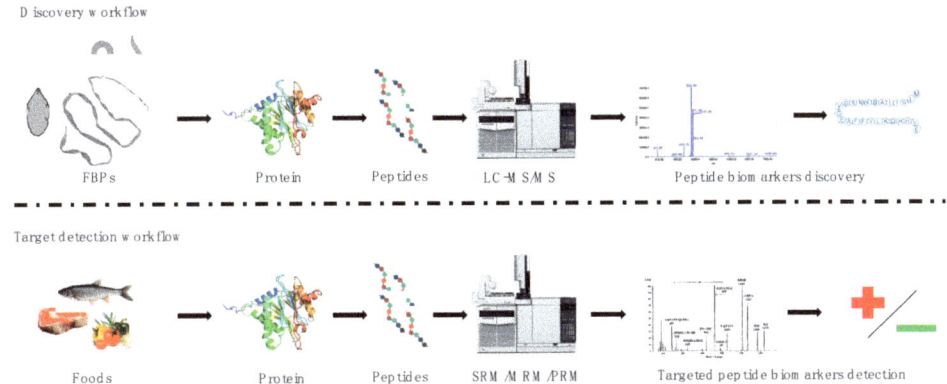

Figure 1. Schematic representation of two proteomic strategies in the studies of foodborne parasites (FBPs): discovery and targeted workflows. Red plus—positive detection (sample contaminated); green minus—negative detection (sample free); LC-MS/MS—liquid chromatography tandem mass spectrometry; SRM/MRM/PRM—selected/multiple/parallel—reaction monitoring.

The aim of this paper is to present a review of the proteomics methods applied to (i) discovery phase —the studies of FBPs with particular attention to identifying and characterizing new targets for treatment and diagnosis, and to (ii) targeted detection phase —selected FBPs detection in food products.

2. Discovery Approach—Description of the Selected FBPs and the Proteomics Methods Used to Study Them

Priorities in FBP differ at the global and European levels. In this work, we focused on the most important emerging FBPs in Europe according to the ranking for risk management of FBPs created for the recommendation of the Food and Agriculture Organization of the United Nations (FAO) and the World Health Organization (WHO) [36]. Additionally, we divided FBPs according to transmission routes (Table 1). An overview on the systematics of selected parasites is provided in Table 2. A brief description of selected FBPs with emphasis on proteomics methods used to study them is provided below. For more detailed information on the pathogenesis and surveillance of diseases caused by foodborne parasites, please see the individual disease health topic pages and factsheets on the European Centre for Disease Prevention and Control website (https://www.ecdc.europa.eu/en) and additionally on the Centers for Disease Control and Prevention website (https://www.cdc.gov) and the World Health Organization website (https://www.who.int).

Table 1. Ranking of foodborne parasites in terms of their importance and risk for european countries according to the criteria of the World Health Organization (WHO) and Food and Agriculture Organization of the United Nations (FAO). To see detailed multi-criteria decision analyses, see reference [36].

Rank	Foodborne Parasites	Infective Life Stage	Transmission Route
1	*Echinococcus multiocularis*	Eggs	soilborne
2	*Toxoplasma gondii*	Fecal oocyst or tissue cyst (bradyzoites)	soil- and meatborne
3	*Trichinella spiralis*	Larvae in a nurse cell	meatborne
4	*Echinococcus granulosus*	Eggs	soilborne
5	*Cryptosporidium* spp.	Oocysts	waterborne
6	*Trichinella* spp. other than *T. spiralis*	Larvae	meatborne
7	Anisakidae	Larvae	seafood-borne
8	*Giardia lamblia*	Cysts	waterborne
9	*Toxocara* spp.	Eggs	soilborne
10	*Taenia solium*	Eggs/Cysticerci	meatborne
11	Opisthorchiidae	Metacercariae	seafood-borne
12	*Ascaris* spp.	Fertilized eggs	soilborne
13	*Angiostrongylus cantonesis*	Larvae	seafood-borne
14	*Entamoeba histolytica*	Cysts	waterborne
15	*Taenia saginata*	Eggs/Cysticerci	meatborne
16	*Diphyllobothrium* spp.	Plerocercoid larvae	seafood-borne
17	*Fasciola* spp.	Metacercariae	plantborne
18	*Sarcocystis* spp.	Cysts with bradyzoites	meatborne
19	*Trypanosoma cruzi*	Metacyclic trypomastigotes	soilborne
20	*Balantidium coli*	Cysts	soil- and waterborne
21	*Cyclospora cayetanensis*	Sporulated oocysts	waterborne
22	*Trichuris trichiura*	Eggs	soilborne
23	*Paragonimus* spp.	Metacercariae	seafood-borne
24	Heterophyidae	Metacercariae	seafood-borne
25	*Spirometra* spp.	Pro-/Plerocercoid larvae	water- and meatborne

Table 2. Selected FBPs systematics. Taxonomy has been adopted from the National Center for Biotechnology Information (https://www.ncbi.nlm.nih.gov/taxonomy). For the purposes of this work, slected species of FBPs from three main systematic groups (Platyhelminthes, Namatoda, and Protozoa) were listed. Additionally, the common name of the parasite was added if applicable, the name of the disease caused by each FBP, as well as the human organs where the parasites occure. Legend: *—phylum, ˆ—class, "— clade, '—order.

	Systematics			Species/Caused Disease	Common Name/Human Organ Where Occures
Flatworm infection	Platyhelminthes *	Rhabditophora ˆ	Trematoda " Plagiorchiida '	*Fasciola hepatica/ F. gigantica* Fasciolosis	Common liver fluke; liver
				Opisthorchis viverrini/ O. felineus Opisthorchiasis	Southeast Asian/Cat liver fluke; liver
				Paragonimus westermani Paragonimiasis	Oriental lung fluke; lung
				Heterophyidae Heterophyiasis	- ; small intestine
		Cestoda ˆ	Cyclophyllidea '	*Echinococcus granulosus/ E. multilocularis* Echinococcosis	Dog tapeworm/Hydatid worm; liver and other organs
				Taenia saginata/T. solium Taeniasis/Cysticercosis	Beef/Pork tapeworm; small intestine
			Pseudophyllidea '	*Diphyllobothrium latum* Diphyllobothriasis	Broad fish tapeworm; small intestine
				Spirometra spp. Sparganosis	- ; subcutaneous tissues or muscle

Table 2. Cont.

	Systematics			Species/Caused Disease	Common Name/Human Organ Where Occures
Roundworm infection	Nematoda *	Chromadorea ^	Rhabditidia '	*Angiostrongylus cantonensis* Angiostrongyliasis	Rat lungworm; brain and nervous system
			Ascaridida '	*Ascaris lumbricoides* Ascariasis	Large roundworm; small intestine
				Anisakis simplex s.s. /*A. pegreffii* Anisakiasis	Herring worm; gastrointestinal tract
				Toxocara canis/*T. cati* Visceral larva migrans/Toxocariasis	Dog/feline roundworm; eye, liver, lungs etc.
		Enoplea ^	Trichocephalida '	*Trichinella spiralis* Trichinosis	Trichna worm; intestine, muscle and sometimes other organs
				Trichuris trichiura Trichuriasis	Whipworm; large intestine
Protozoan infection	Apicomplexa *		Eucoccidiorida '	*Toxoplasma gondii* Toxoplasmosis	-; brain, eye, lungs, heart, muscle etc.
				Cryptosporidium parvum Cryptosporidiosis	-; intestinal tract
				Sarcocystis spp. Sarcocystosis	-; blood vessels, muscles, intestine
				Cyclospora cayetanensis Cyclosporiasis	-; stomach, small intestine
	Metamonada *		Diplomonadida '	*Giardia lamblia* Giardiasis	-; small intestine
	Amoebozoa *		Amoebida '	*Entamoeba histolytica* Amoebiasis	-; large intestine and other organs
	Euglenozoa *		Kinetoplastida '	*Trypanosoma cruzi* Chagas disease	-; heart, oesophagus, colon, nervous system
	Ciliophora *		Heterotrichida '	*Balantidium coli* Balantidiasis	-; cecum and colon

2.1. Waterborne Parasitic Species

Water is the main habitat for many lifestages of parasites. These stages can contaminate food products or directly infect humans via the drinking of infected water. Among waterborne parasitic infection, according to the European ranking [36], the most important parasites are *Cryptosporidium* spp. (fifth/25), *Giardia lamblia* (eight/25), *Entamoeba histolytica* (14th/25), *Cyclospora cayetanensis* (21st/25), and *Spirometra* spp. (25nd/25).

2.1.1. *Cryptosporidium* spp.

Cryptosporidium spp. is one of the main causes of human diarrhoeal diseases and, with *Giardia lamblia*, is a major cause of protozoan waterborne diseases [3,37]. In addition to diarrhea, the general symptoms associated with cryptosporidiosis include nausea, vomiting, loss of appetite, and cramps. Cryptosporidiosis is having a clear link with impaired cognitive and functional development in children in developing countries [38]. The lifecycle of *Cryptosporidium* spp. is completed within a single host. In immune-competent individuals, the symptoms of cryptosporidiosis are often self-limiting, but may be chronic when the immune system is compromised, such as in children under 5 years of age or cancer patients [39]. There is currently no effective medication for cryptosporidiosis [40], and so far, nitazoxanide is the only drug accepted by U.S. Food and Drug Administration. Despite that, as mentioned earlier, it is still ineffective, most of all for immunosupressive patients, such as people living with HIV [41]. Between 1984 and 2017, waterborne disease outbreaks, as well as infections through unpasteurized milk and dairy products, and through handling animals, and infections by using recreational waters were caused by 25 outbreaks of cryptosporidiosis [3,7,37].

The use of proteomics techniques was important during the studies of *Cryptosporidium* spp. Sanderson et al. [42] conducted an extensive analysis of the proteome of excysted *C. parvum* sporozoites. Three independent proteomics methods were used to maximize the coverage of the proteome: (i) 2-DE LC-MS/MS; (ii) 1-DE LC-MS/MS; and (iii) multi-dimensional protein identification technology (MudPIT) analysis, in which trypsin-digested peptides were separated by multi-dimensional LC and then subjected to MS/MS. Over than 4800 protein spectra have been identified. These proteins represent 1237 non-redundant proteins, what is one third of the entire proteome of *C. parvum*. For example, Siddiki and Wastling [43], used mass spectrometry-based basic local alignment search tool (MS BLAST) to identify *C. parvum* proteins from frozen sporozoite pellets isolated from lamb feces. They separated the total protein by one-dimensional sodium dodecyl sulphate–polyacrylamide gel electrophoresis (1D-SDS-PAGE) and analyzed by two-dimensional nano-liquid chromatography-tandem mass spectrometry (2D-n-LC MS/MS). Using this method, the authors found 84 proteins specific for *C. parvum*, one third of which were previously hypothetical. In another study, Snelling et al. [44] also used MS, but they not only tried to analyze the proteome of Cryptosporidium, but also aimed to determine proteins that were differentially regulated in the excysted sporozoites compared to the non-excysted. Their proteomic analysis shown the expression of 26 proteins, which were significantly modulated after excystation. Interestingly, 3 of the proteins were specific for apicomplexan, and 5 were specific for Cryptosporidium. The authors proposed that, all identified proteins may be involved in pathogenesis. However, it remains to be determined whether Cryptosporidium causes the same response when it comes into contact with the host or is internalized. Beside these "omic" studies, using an in silico approach, the novel drug target (protein) was described and characterized using predicted proteome and bioinformatics methods. However, an in vitro or in vivo study is needed to confirm the above proposition [45].

At present, we hardly understand which proteins interact with the parasite/host, because the data retrieved from the host-parasite protein-protein interaction from the Cryptosporidium-infected hosts is very limited.

2.1.2. Giardia lamblia

Giardia lamblia (*G. lamblia*) (also known as *Giardia intestinalis* and *Giardia duodenalis*) is a single-celled protozoan parasite that can infect the small intestines of humans and animals. Giardiasis occurs globally across socioeconomic boundaries but is mainly endemic in developing countries and particularly within young children [22]. Main symptoms of giardiasis are diarrhea, epigastric pain, nausea, vomiting and weight loss, and they appear 6–15 days after infection and are more severe on young children and individuals with malnutrition or immunodeficiency. Giardiasis is usually treated with metronidazole or other nitroimidazoles [46].

Proteomics were widely used in the studies on *G. intestinalis*. Originally, the secretome of *G. intestinalis* after in vitro co-incubation with human intestine cell lines (Caco-2 cells or HT-29) was analyzed using a 2D gel-based approach, with proteins identified using matrix-assisted laser desorption ionization time-of-flight mass spectrometry (MALDI-TOF MS) [47]. This experiment identified three metabolic *G. intestinalis* proteins (arginine deiminase, ornithine carbamoyl transferase, and enolase) and two human proteins (enolase and HSP70). Then, the Giardia secretome was further and more deeply explored in subsequent studies [48,49]. Ma'ayeh et al. [49] analyzed the *Giardia* secretome during host–parasite interaction between differentiated Caco-2 cells and isolates from both *G. intestinalis* human-infective assemblage (WB and GS) using LC-MS/MS. Dubourg et al. [48] also identified steady-state, axenic-secreted proteins in *G. intestinalis* (WB and GS isolates) proteins. The label-free method was used for protein quantification; the intensity-based absolute quantification (iBAQ) approach, which calculates the sum of parent ion intensities of the peptides identified in each protein. Secreted proteins were identified based on ratios between protein abundance of proteins analyzed in whole trophozoite lysate compared to protein abundance from culture supernatants. Higher supernatant to lysate ratios of protein abundance were considered indicative of enrichment due to active parasite secretion and used to eliminate cytoplasmic contaminants derived from parasite lysis [48]. The findings from these studies supports observations that secreted proteases are important virulence factors in breaking down host gut barriers and modulating host immune responses. Recently it was demonstrated that *Giardia* trophozoites release microvesicles (MV) which play key role in proliferation and parasite-host interaction. The protein repertoire of peripheral vesicles (PVs) and encystation-specific vesicles (ESVs) has been described [50]. To describe the first protein composition of ESVs and PVs, a novel strategy combining flow cytometry-based organelle sorting with in silico filtration of mass spectrometry data was used (SDS-PAGE and LC ESI-MS/MS) [50]. The protein composition of MV was also analyzed using MS from both trophozoites and cysts, with 11 and 80 proteins identified in MV from each lifestage, respectively [51].

Recently, in order to expand the variety of vaccine candidate antigens, Davids et al. [52] considered that surface proteins can be a rich source of such antigens. In their study, the trophozoites reacted with non-membrane-permeant NHS biotin. The biotin-labeled and unlabeled controls were evaluated by immunofluorescence. Differential interference contrast microscopy was used to compare cells. All cell lysates were prepared in parallel and analyzed either directly or immunoprecipitated using streptavidin-agarose beads. The precipitated proteins were separated by SDS-PAGE, and gels were visualized by Coomassie staining. The biotinylated proteins were detected by immunoblotting with an anti-biotin antibody. Immunoprecipitants were then digested and analyzed by LC-MS/MS [52].

The latest research on the surface proteome of *G. intestinalis* used shotgun mass spectrometry to analyze the proteomes of trophozoites of the three different strains. This allowed the identification of 2368 proteins, among which, using monoclonal antibodies, the variant-specific surface proteins were identified [53].

Accordingly, the proteomics helped to conclude that there is an extensive host-parasite crosstalk during *Giardia* infections of different human cell lines via both secretome and direct interactions.

2.1.3. Cyclospora cayetanensis

Cyclospora cayetanensis (*C. cayetanensis*) is an apicomplexan, coccidian protozoan parasite, which in 1994 was described as a causative agent of human cyclosporiasis—a self-limiting diarrheal disease, the symptoms of which are also fatigue, loss of appetite, nausea, and vomiting [54]. Infection occurs through the ingestion of contaminated water or agricultural products (raspberries, basil, and coriander) [55,56]. Since the spore formation time after fecal shedding is longer (at least seven days), the possibility of infection between people is less. Cyclosporiasis is endemic in Nepal, Peru, and Haiti [57,58].

Little is known about the proteome of *C. cayetanensis*. Cinar et al. [59] sequenced the *C. cayetanensis* genomic DNA extracted from clinical stool samples, annotated, and used the sequence as a reference for proteome prediction. Therefore, they validated the quality of such reference by comparing the predicted proteome of related parasites (including Cyclospora and Babesia). The analysis showed 29 core apicomplexan proteins found in most apicomplexans [59]. A similar method was recently used by Liu et al. [60], who proved that the *C. cayetanensis* genome may encode as many as 7457 proteins. Among them, 538 proteins had signal peptides (of which 105 target the apicoplast), 1247 had one or more transmembrane regions, and 225 had a GPI anchor attachment site. These data are similar to those in *Eimeria tenella* and *Toxoplasma gondii* [60].

2.1.4. Entamoeba histolytica

Amoebiasis is a disease caused by the protozoan parasite *Entamoeba histolytica* (*E. histolytica*). When a person swallows something contaminated with *E. histolytica* cysts (water or food products), infection may occur [61]. Only about 10% to 20% of people who are infected with *E. histolytica* will become ill due to the infection. Symptoms include loose stools, stomach pain, and stomach cramps. Amoebic dysentery is a serious form of amoebiasis and is related to stomach pain, bloody stools and fever. *E. histolytica* rarely invades the liver and forms an abscess [62].

Most of the proteomic publications have focused on analyzing the expression profiles of trophozoites in different organelles under numerous conditions [61]. By 2-DE followed by MS, whole extracts have been analyzed, including both soluble and insoluble proteins, cytoskeletal, membrane, and signaling-associated proteins [63,64]. Other teams focused on investigating phagosome composition and proteins associated with the phagocytosis process [65], the cell surface protein profile [66], and a nuclear and cytoplasmic proteomes of trophozoites [67]. In a more recent study, the membrane proteome of *E. histolytica* was described [68], and the excretory-secretory proteins were identified [69]. The components of trophozoite ER and Golgi apparatus were characterized as well [70]. In addition, by comparing the proteome of *E. histolytica* with related non-pathogen amoeba *E. dispar* it was possible to identify proteins related to virulence and pathogenicity [71–74]. The latest studies focused not only on the trophozoite stage of *E. histolytica* but also on the cysts and cyst-like structures gained from trophozoites [61,75]. Moreover, *E. histolytica* exposed to serum isolated from a person with amoebiasis, induces cell polarization by activating signal transduction pathways and cytoskeletal components. This process results in the formation of a protruding pseudopod at the front of the cell and a retracted uropod at the rear. Marquay Markiewicz et al. [76], using LC-MS/MS, showed the proteomic composition of the uropod fractions.

The proteins identified in presented studies that may be recognized by the immune system and/or released into the circulatory system during amoebiasis, and may be a detectable biomarker of the disease.

2.1.5. *Spirometra* spp.

The procercoid larvae or plerocercoid larvae of *Spirometra* tapeworms are the cause of sparganosis in humans [77]. Humans are an intermediate host for the parasite, who acquire sparganosis most often by drinking water contaminated with infected copepods (intermediate host) or consuming the meat of an undercooked second intermediate host (fish, reptiles, amphibians). Once ingested by a human, the larvae undergo visceral migration and can end up in many tissues, where they grow [78]. Depending on the final location of the parasite, migrating sparganum can cause various symptoms. It may be located in almost every part of human body, including subcutaneous tissue, breasts, orbits, urinary tract, lungs, pleural cavity, abdominal viscera and central nervous system [78]. The migration of subcutaneous tissue is commonly painless. However, when the parasite settles in the one of the elements of central nervous system, like brain or spine, a variety of neurological symptoms may occur, including weakness, headaches, seizures and abnormal skin sensations such as numbness or tingling. If the inner ear is affected, the patient may experience dizziness or may even become deaf for a short time [77].

Proteomic methods have been used to characterize *Spirometra* tapeworms. 2-DE was used to describe the protein expression differences between three different stages of *S. erinacei*, the plerocercoid larvae, eight-day-old juveniles, and adults [79]. The specific or highly expressed proteins in juvenile worms were analyzed by MALDI-TOF MS/MS. The proteome profile of larvae showed fewer protein spots than juveniles or adults, and juveniles and adults showed similar protein expression profiles. Eight juvenile-specific proteins and five juvenile up-regulated proteins were identified and their functions were determined [79]. Immunoproteomic analyses of *S. erinaceieuropaei* and *S. mansoni* have also been performed [80,81]. Both studies used 2-DE and Western blot probed with sera from infected mice. Protein spots which showed immune response were characterized by MALDI-TOF/TOF-MS. The recent study on *S. erinaceieuropaei* sparanga described site-specific phosphoproteome, with the purpose of describing the global phosphorylation status of spargana [82]. A total of 1758 spargana proteins were identified, where 3228 phosphopeptides and 3461 phosphorylation sites were described among. This dataset provides a valuable data repository for future research on the metabolic pathways of this important zoonotic parasite.

2.2. Soil- and Plant-Borne Parasitic Species

Infective developmental stages of parasites are spread with fecal-contaminated soil and may contaminate food products like editable aquatic plants (e.g., watercress, algae), vegetables, fruits, and fruit juices. Among soil- and plant-borne parasitic infection, according the European ranking [36], the most important parasites are *Echinococcus multiocularis* (first/25), *E. granulosus* (4th/25), *Toxocara* spp. (ninth/25), *Ascaris* spp. (12th/25), *Fasciola* spp. (17th/25), *Trypanosoma cruzi* (19th/25), and *Trichuris trichiura* (22nd/25).

2.2.1. *Echinococcus multiocularis* and *E. granulosus*

Six species of genus *Echinococcus* have been described, of which two are of public health importance in Europe: *E. granulosus* (the causative agent of cystic echinococcosis—CE) and *E. multiocularis* (the causative agent of alveolar echinococcosis—AE) [36]. The infection is caused by the accidental consumption of fruit and vegetables contaminated with parasite eggs shed by a carnivore final host [3]. For both these species, humans are accidental intermediate hosts [3]. In humans, the liver is the most common site for cystic and alveolar echinococcosis [83].

In CE, cysts grow slowly (1–5 cm in diameter per year). It may take many years to show any symptoms, usually due to organ dysfunction in which the cyst grows. If a cysts ruptures, the sudden release of its contents can cause sensitization ranging from mild to severe anaphylactic shock [84]. The immunodiagnostic techniques coupled with anamnesis, and radiological imaging are used for diagnostic purposes of echinococcosis where, the steps for diagnosing AE in humans are the same as those for CE [85,86]. There are several main treatment options, including surgery, puncture aspiration injection reaspiration, and chemotherapy. For asymptomatic individuals, consideration of a "waiting and watching" approach with the supervision of the patient is recommended [83,87].

The proteomes of *E. granulosus* and *E. multiocularis* are still not well described, but there are some reports of proteomic studies on those parasites. In 2003, for the first time the proteomic analysis of *E. granulosus* protoscoleces by 2-DE and peptide mass fingerprinting (PMF) was performed [88]. Host serum proteins (especially albumin and globulin) were highly concentrated in the samples, what caused horizontal streaks on the hydatid fluid 2-DE gels. Even when parasite hydatid fluid-enriched fraction was prepared, large amounts of bovine serum albumin and globulins still made it complicated for 2-DE to detect parasite-specific proteins. A few more studies describing protoscolex proteome using 2-DE were performed, some of which were followed by MS analysis [89–93]. Then, Monteiro et al. [94] used LC-ESI-Q-TOF MS/MS to analyze protoscolex and the hydatid cystic fluid of *E. granulosus*. Moreover, Longuespée et al. [95] in order to describe the proteomics model of CE in the liver, used the latest laser microdissection-based proteomics and MALDI-MS workflow. This study demonstrated specific markers of a parasitic cyst in the liver. The comparison of *E. multilocularis* and *E. granulosus* hydatid fluid protein composition was also done using LC-ESI-MS/MS and provides explanation of specialized host–parasite interactions [96]. The proteome of an adult stage of *E. granulosus* was also described [97]. Moreover, the extracellular vesicles derived from *E. granulosus* and their protein composition was also presented [98].

2.2.2. *Toxocara* spp.

Toxocariasis is caused by the transmission of *Toxocara* species from carnivores (canines or cats) to humans. The most widespread and very epidemiologically important species in the world, *T. canis* can infect a large group of canines, like dogs, foxes, wolves, jackals, and coyotes, while *T. cati* can infect cats [99]. The human can accidentally ingest *Toxocara* eggs containing infectious third-stage larvae (L3) from contaminated food, environment (soil or sand) and/or water. The L3 larvae hatch the egg, and migrate through the wall of the host's small intestine, and then can get through the circulatory system to the variety of organs, including liver, lungs, central nervous system and/or muscle tissue [100]. Most infections are asymptomatic, and since clinical investigations and/or diagnostic tests are not usually performed, human diseases may go unnoticed but sometimes larvae cause immune and inflammatory reactions, resulting in symptoms including fever, headache, cough, and pain in the abdomen or limbs. There is currently no vaccine against toxocariasis. Human chemotherapy differs depending on the symptoms and location of the larva but usually is limited to albendazole or mebendazole administered with anti-inflammatory corticosteroids [99].

Few studies using proteomics methods have been conducted on *Toxocara* spp. In this review, we focused only on describing *T. canis*. The *T. canis* genome contains at least 18,596 protein-coding genes, and their predicted products include at least 373 peptidases, 458 kinases, 408 phosphatases, 273 receptors, and 530 transporters and channels. In addition, the secretory proteome (870 molecules) of *T. canis* is rich in proteases that are probably involved in the penetration and degradation of host tissues, and rich in molecules that are recommended to inhibit the host's immune response [101]. Analysis of excretory-secretory products and larval extract using 1D-SDS-PAGE-LC-MS/MS has been done by da Silva et al. [102], who identified 646 proteins (582 somatic and 64 excretory-secretory), among which, many may play a role in parasite-host interactions, as well as in regulating parasite metabolism and survival. A similar approach was used by Sperotto et al. [103], who identified 19 proteins. According to the classification using the signal peptide predicted by SignalP [104], 7 of the identified proteins were located outside the cell, 10 have cytoplasmic or nuclear localization, and the subcellular localization of the two remain proteins was unknown [103].

This advancement in *Toxocara* proteomics has brought hope to medicine and veterinary medicine, especially in the areas of better diagnostic tools, effective vaccines or drugs.

2.2.3. *Ascaris* spp.

Ascaris lumbricoides (*A. lumbricoides*) and *A. suum* are infecting humans and pigs, respectively [105]. *A. lumbricoides*, as a one of the most common parasites in the world is infecting 1.2 billion people worldwide [106]. In human and pig hosts, the migration route of larvae is similar. After ingesting infectious ova, L3 larvae covered by the L2 cuticle hatch in the small intestine and migrate to the caecum and proximal colon, where they penetrate the mucosa. The larvae then migrate via the portal vein to reach the liver, where the L2 cuticle is shed. The larvae then migrate through the portal vein to reach the liver. After migrating in the liver, the larvae enter the lungs 6-8 days post infection (p.i.), then penetrate the alveolar space and move to the pharynx, where they are swallowed, what is causing larvae to return to the small intestine on day 8–10, where larvae molt to L4 development stage. Larvae mature and reach sexual maturity on day 24 p.i. and during that time is molting last (L5) [105]. Due to the hepato-tracheal migration, the infection with *A. lumbricoides* may cause pulmonary and intestinal symptoms such as persistent sore throat, dyspnoea, sometimes coughing up blood, abdominal pain, nausea, vomiting, and diarrhea. In addition, eosinophilic pneumonia (pulmonary ascariasis) or obstruction of the intestinal lumen in the case of intestinal ascariasis may occur in case of severe infection. Ascariasis is treated pharmacologically by administering albendazole or mebendazole in a single dose [107].

Despite the public health importance impact, both parasites proteomes' are poorly described. In the latest paper, Xu et al. [108], described the use of 2-DE coupled with MALDI-TOF/TOF MS to compare proteomes of adult female *A. lumbricoides* and *A. suum*. In the six gels examined (three gels for each parasite species), more than 630 and 750 protein spots were repeatedly found. After comparing the 2-DE proteomes of *A. lumbricoides* and *A. suum*, it was found that the protein profiles of the two species were very similar, with almost no differentially modulated proteins. The protein expression profiles determined by 2-DE coupled with MALDI-TOF/TOF MS method were about three times higher than those obtained with 2-DE [109]. Analysis of only head ends of 10 immature *A. lumbricoides* and *A. suum* using MALDI-TOF MS was also performed, but only to describe protein profiles (characteristic protein peaks) of those species [110]. Most of the proteomic studies were performed on *A. suum*. Proteomic analysis of the excretory–secretory products from larval stages (L3-egg, L3-lung, and L4) of *A. suum* by LC-MS/MS revealed high abundance of glycosyl hydrolases. Another, immunoproteomic approach using 2-DE-MALDI-TOF MS and sera from pigs with ascariasis let to identify 24 immunoreactive proteins [111]. Most of them (23/24) were determined to be related to the survival mechanisms of parasites, involving functions connected with energy production (12 proteins) and redox processes (5 proteins). These results might help to find effective chemotherapeutic targets for porcine ascariasis [111]. Different strategy was used to characterize the protein composition of perienteric fluid (PE), uterine fluid (UF), and total excretory/secretory products (ESP) from this parasite. Chehayeb et al. [112] used SDS-PAGE combined with LC-MS/MS to identify 175, 308, and 274 proteins in ESP, PE, and UF, respectively. The ultra-performance liquid chromatography coupled to nano-electrospray tandem mass spectrometry (UPLC-nanoESI MS/MS) let to identify 268 proteins of extracellular vesicles isolated from *A. suum* by ultracentrifugation. To date, to our knowledge, it is the most comprehensive analysis of protein composition of *A. suum* extracellular vesicles [113].

2.2.4. *Fasciola* spp.

Six different species of plant-borne trematodes are known to affect humans: *Fasciola hepatica*, *F. gigantica*, *Fasciolopsis buski* (Fasciolidae), *Gastrodiscoides hominis* (Gastrodicidae), *Watsonius watsoni*, and *Fischoederus elongates* (Paramphistomidae). Whereas *G. hominis*, *F. buski*, *W. watsoni*, and *F. elongates* are intestinal, the *F. hepatica* and *F. gigantica* are hepatic trematodes. In the present section, we will focus only on the members of Fasciola. Fascioliasis is caused by two species of liver fluke—*F. hepatica* and *F. gigantica*. Fascioliasis affects domestic animals, as well as humans [114]. *F. hepatica* is a cosmopolitan species because it has the ability to infect many different species, and the intermediate snail host has the ability to adapt to various ecological niches [115]. Due to the reduced ability of aquatic snail intermediate hosts to invade new niches, the distribution of *F. gigantica* is more limited, usually to tropical regions of Asia and Africa [116,117]. Outbreaks of human infections are always related to local animal fascioliasis cases [115,117]. It is estimated that between 2.4 and 17 million people are currently infected and 91 million are at risk of infection [118]. Infection is usually spread by various aquatic plants, such as watercress, algae or tortora, on which the metacercaria have settled and are then consumed [114]. Farm management practices and growing aquatic plants in greenhouses have reduced the number of human infection cases in industrialized areas, but still in some developing countries, wild aquatic plants or plants grown in fields, where infected animals can roam freely, they become a threat to humans. In addition, metacercaria can be found floating in the water, so people can get the infection by drinking water [3,116,117]. Clinical symptoms of an acute fascioliasis are abdominal pain, indigestion, weight loss, mild fever. Other gastrointestinal symptoms result from the migration of the young flukes through the liver, which also always results in hepatomegaly [30].

Proteomics has been widely used in the studies on *F. hepatica* excretory–secretory (ES) proteins [119]. Early proteomic studies of *F. hepatica* used radiometabolic markers to distinguish protein profiles at different developmental stages [120,121]. Isoelectric focusing and densitometry were also carried out to characterize the ES proteins secreted by flukes parasitizing diverse mammals [122,123]. Jefferies et al. [124,125] improved this analysis using 2-DE. These studies characterized a range of different glutathione S-transferases, fatty acid binding proteins, superoxide dismutase, peroxiredoxin, and cathepsin L-proteases, which have been further analyzed using more advanced proteomics techniques [126–134]. Proteomic methods (2-DE-LC-MS/MS) were also adjusted to describe the mechanism of action of anthelminthic drug triclabendazole on *F. hepatica* [135], as well as to discern protein signatures of this parasite's susceptible and putatively resistant to triclabendazole [136]. One challenge in the proteomics of *F. hepatica* was to characterize the proteomes of early developmental and migratory stages because of their small size. Moxon et al. proved that eggs have significantly different proteomes from the other life stages of *F. hepatica* [137]. DiMaggio et al. [134], using gel-free proteomic techniques (shotgun proteomics), performed an extensive analysis of the proteins secreted by an adult *F. hepatica* and newly excysted juveniles (NEJ; 48 h post-excystment) and compared these with the somatic proteome of the NEJ 48 h. This study identified 202 proteins in the adult secretory group, 90 proteins in the NEJ 48 h secretory group, and 575 proteins in the NEJ 48 h somatic proteome.

The key to survival in the host environment is the parasite surface that can change quickly to prevent host immune cells from attacking. Proteomic characterization (nanoUPLC–ESI–qTOF–MS and MALDI-TOF-MS) of the *F. hepatica* tegument was also performed [138,139].

2.2.5. *Trypanosoma cruzi*

Trypanosoma cruzi (*T. cruzi*) is a species of parasitic euglenoids, which are transmitted to mammals by insects from the subfamily of the Reduviidae, the hematophagous insect triatomine ("assassin bug," "cone-nose bug," and "kissing bug") [140]. The natural habitat of bugs are the burrows and nests of animals. *T. cruzi* reservoirs are armadillos and opossums and, less frequently, rodents, monkeys, dogs, cats, and cattle [141]. Invasive forms for mammals, including humans (metacyclic trypomastigotes), are excreted onto the skin along with feces of bugs during blood sucking. Invasive forms reach the wounds at the site of the bug bites and skin scratches and also through the mucous membranes and the conjunctiva (rubbing the eye with the hand). In the host's cells, parasites undergo part of the life cycle (from metacyclic trypomastigotes to amastigotes and then trypomastigotes). When infected cells are lysed, parasites are released, then enter subsequent cells and infect them. If the trypomastigote enters the gut of a bug during blood sucking from an infected host, it will complete the life cycle [142]. The disease caused by *T. cruzi* in humans is called Chagas disease (also known as American trypanosomiasis) [143]. Chagas disease has two stages: the acute stage, which develops 1–2 two weeks after the insect bite, and the chronic stage, which develops for many years. Usually asymptomatic is the acute phase [143,144]. In the chronic stage, symptoms include fever, general malaise, headache, hepatomegaly, splenomegaly, and swollen lymph nodes. It is rare for people to have nodules swelling at the site of infection. If it is on the eyelid, it is called "Romaña's sign" and if it is on the skin, it is called "chagoma" [144]. In rare cases, an infected person will develop a severe acute illness, which may cause life-threatening fluid accumulation around the heart or cause inflammation of the heart or brain. The acute phase usually lasts 4 to 8 weeks and will subside without medication [143].

Proteomics has been widely used in the studies on *T. cruzi*. Large-scale comparative proteomics studies have shown that metacyclic form had the highest number of proteins expressed, followed by amastigotes, epimastigotes, and trypomastigotes [145]. Many other studies showed stage-specific proteins. The shotgun proteomics of the blood trypomastigote stage was used to described the main classes of proteins present in this stage [146]. Comparison of the proteome of blood trypomastigote with that derived from the tissue culture or metacyclic trypomastigote shows that more than 2200 proteins are unique to the blood trypomastigote stage and participate in various cellular processes [146]. Proteomic analysis of the trypomastigote identified more than 1400 proteins, of which nearly 14% are surface proteins anchored by glycophosphatidylinositol (GPI), which might be involved in host cell invasion and immune escape [147]. A study showed that the difference between the protein expressions of the epimastigotes and trypomastigotes was more than 50%. The study also determined that some protein isoforms are involved in metacyclogenesis [148]. Metacyclogenesis, the process of transforming procyclic promastigotes into highly infective metacyclic promastigotes, was also investigated using proteomic tools. Large-scale proteomics research has pointed out major differences in proteins related to oxidative stress, translation, and metabolic pathways related to proteins, lipids and carbohydrates [149]. Quantitation of phosphorylated proteins in the same study showed that there are more than 7000 phosphorylation sites, of which 260 are under different regulation, including some potential drug targets, e.g., sterol biosynthesis enzymes. Next study indicated that during periods of nutritional stress, many proteins are phosphorylated, which may trigger metacyclogenesis [150]. Similarly, the proteomic comparison of the exponential phase and stationary phase of the epimastigotes quantified more than 3000 proteins [151]. The transition from trypomastigotes to amastigotes (amastigogenesis) was described by quantitative proteomics and phosphoproteomics [152,153].

The surface proteomes of *T. cruzi* was also studied. Comparison of surface proteomes at different stages showed that most of the proteins are expressed in more than one stage, but several are specific for particular stage [153]. Another study showed membrane-derived proteins can participate in invasion, adhesion, cell signal transduction, and modify the host's immune response. A new family of surface membrane proteins called TcSMPs (*T. cruzi* surface membrane proteins) that is conserved among different *T. cruzi* lineages has been characterized [154].

2.2.6. *Trichuris trichiura*

Trichuris trichiura (*T. trichiura*) (whipworm) is a nematode that causes trichuriasis in humans. The larvae infects human cecum and colon [155]. Ingestion of embryonated eggs from the external environment can cause infection. After hatching, the larvae emerge from the polar egg and establish an infection in the epithelium of the cecum and colonic Lieberkühn crypts. Following the characteristic four molts, the dioecious adult parasites develop unobstructed (rate depends on the host), mate and hatch unembryonated eggs, which are expelled into the environment through feces. [155]. Trichuriasis is more common in warm climates. If the infected person defecates outdoors, or if untreated human feces is used as fertilizer, the eggs are deposited on the soil and they can mature to the infection stage. Ingestion of these eggs "can happen when hands or fingers that have contaminated dirt on them are put in the mouth or by consuming vegetables or fruits that have not been carefully cooked, washed or peeled" [156].

The proteomic studies were not conducted often on *T. trichiura*. In 1995, the 2D-SDS-PAGE electrophoresis technique was used to describe the ability of excretory/secretory proteins of *T. trichiura* adult worms recovered from the human, to provoke an immune response [157]. Latest proteomic analysis of *T. trichiura* egg extracts using LC-MS/MS revealed potential immunomodulatory and diagnostic targets [158]. Most of all the other proteomic studies, to our knowledge, were conducted on *T. muris*, which has been used for over 50 years as a model for *T. trichiura* [159]. The best-described issue using proteomics methods is the extracellular vesicles' protein composition. *T. muris* proteins from the vesicular component were analyzed by LC-MS/MS in several studies, and potential immunogenic proteins and new insights into parasite–host communication were described [160–162].

2.3. *Meat-Borne Parasitic Species*

Humans get infected by many FBPs by eating uncooked or raw meat infected with development stages of these parasites. Among meat-borne parasitic infection, according the European ranking [36], the most important parasites are: *Toxoplasma gondii* (second/25), *Trichinella spiralis* (third/25), *Trichinella* spp. other than *T. spiralis* (sixth/25), *Taenia solium* (10th/25), *Taenia saginata* (15th/25), and *Sarcocystis* spp. (18th/25). In this section, since there are more proteomics studies from these species, we will only focus on selected parasites.

2.3.1. *Toxoplasma gondii*

Toxoplasma gondii (*T. gondii*) is a protozoan parasite that, like *Cryptosporidium* spp., belongs to the phylum Apicomplexa. The parasite has a complex life cycle for which usually domestic cats are the definitive hosts [163]. Intermediate hosts are all warm-blooded animals, including livestock and humans. Infected cats excrete oocysts in the feces. If they are ingested after spore formation, they will infect the intermediate host and develop into rapidly reproducing tachyzoites, which are spread throughout the body [7]. Future mothers are particularly vulnerable, because tachyzoites can cross the placenta and infect the fetus. After tachyzoites are located in the muscle tissue and central nervous system, they transform into tissue cysts (bradyzoites). Food-borne Toxoplasma infection can be obtained by ingesting tissue cysts in uncooked or raw meat, or ingesting oocysts by eating contaminated vegetables or drinking water [7]. In pregnant women, toxoplasmosis is generally considered a serious health problem, which can transmit infection to fetuses or newborns, as well as in people with weakened immune systems. In adults with strong immunity, the infection is usually asymptomatic [3]. Nevertheless, recently published research have shown that almost all cases of ocular toxoplasmosis is caused by acquired diseases, which means that prevention should not only target pregnant women and people with weakened immune functions, but also the general population [164]. Proper meat cooking, as well as freezing for the appropriate time is the best method known to kill toxoplasma cysts [3].

To our knowledge, *T. gondii* is one of the best proteomic-studied FBPs. Before there was a lot of genomic information about the parasite, the Toxoplasma gondii protein was identified by MS, which depends on the use of NCBI and the limited EST database to identify the protein [165,166]. With the development of an annotated genome for *T. gondii*, it was possible to search efficiently in the MS data against thousands of *T. gondii* protein sequences [167]. The tachyzoite has been the main focus of proteomic studies of *T. gondii*, although some data have now been published on the other life cycle stages. Xia et al. [168] published the results of the first multi-platform (1-DE LC-MS/MS, 2-DE LC-MS/MS and MudPIT) proteomic analysis of *Toxoplasma* tachyzoites, identifying nearly one-third of the entire predicted proteome of *T. gondii*. In another study, Dybas et al. [169] using 1-DE LC-MS/MS identified 2477 gene-coding regions with 6438 possible alternative gene predictions—approximately one third of the *T. gondii* proteome. The proteomics investigation found that compared with any known species (including other Apicomplexan), there are 609 unique proteins of Toxoplasma gondii [169]. The proteomic profiles of different genotypes of *T. gondii* tachyzoites using 2-DE difference gel electrophoresis (DIGE) combined with MALDI-TOF MS were also investigated [170]. A different approach, focused on the antigenicity of soluble tachyzoite antigen (STAg), led to the identification of 1227 proteins of *T. gondii* STAg [171]. Through MS analysis, 426 proteins were identified among the 1227 isolated protein spots. A proteogenomic approach has allowed Krishna et al. [172] to reanalyze many published data sets of *T. gondii* and generate new high-throughput MS/MS data sets. Four different techniques (1-DE, 1-DE of soluble and insoluble fractions (1DE SFIF), 2-DE, and MudPIT) were used and obtained samples were analyzed on an LC-MS/MS. The MS data were searched against the protein database assembled from two different sources: the official gene models and predicted gene models supported by RNA-Seq evidences [172]. With use of this proteogenomic approach the identification of 30,494 peptide sequences and 2921 proteins for *T. gondii* was performed. In addition to the tachyzoite stage, oocysts which are highly resistant to the environmental conditions were also studied by global proteomics methods. Fritz et al. [173] have characterized the proteome of the wall and sporocyst/sporozoite fractions of mature, sporulated oocysts using the 1-DE LC-MS/MS approach. A total of 1021 Toxoplasma proteins were identified in the sporocyst/sporozoite fraction and 226 proteins were identified in the oocyst wall part. Importantly, 172 proteins were identified as not reported in other *Toxoplasma* proteomic evaluations. Moreover, the application of isotope tags for relative and absolute quantification (iTRAQ) coupled with 2-DE LC–MS/MS to investigate the proteome of oocysts during sporulation, let to describe 2095 proteins where 587 were identified as differentially regulated (sporulated and non-sporulated oocysts) [174].

Excretory-secretory proteins were also investigated in *T. gondii*. Zhou et al. [175] have applied proteomics techniques to analyze a large number of freely released Toxoplasma secretory proteins by using 2-DE and MudPIT. Another group using LC-MS/MS identified excretory-secretory proteins from the RH strain of *T. gondii* [176]. A total of 34 proteins were identified and their abundance was estimated by spectral counting method. Among them, eight microparticle proteins (MICs), two species of rhoptry proteins (ROPs) and six dense granular proteins (GRAs) were identified [176].

The most comprehensive description of the proteomic organization of a *T. gondii* cell (tachyzoite) was recently presented by Barylyuk et al. [177] by applying relatively new proteomic method of subcellular localization of thousands of proteins per experiment by isotope tagging (hyperLOPIT). The hyperLOPIT method utilizes a unique abundance distribution map, which is formed during the organelles and subcellular structures biochemical fractionation, e.g., density gradient centrifugation.

Proteins showing similar abundance distribution characteristics through these fractions are assigned to proper subcellular structures [178,179]. In each of three experiment replicates, Barylyuk et al. identified over 4100 proteins across all 10 fractions representing subcellular compartments of *T. gondii* trachyzoite. In addition, these three data sets have a total of 3832 proteins, which can provide complete abundance distribution overview information of 30 fractions. Using the hyperLOPIT approach, Barylyuk et al. assigned thousands of proteins to their subcellular niches [177].

2.3.2. *Trichinella* spp.

Trichinella genus is one of the most widespread group of parasitic nematodes in the world. With the exception of Antarctica, Trichinella infections have been detected in domestic and wild animals on all continents [3]. Not long ago, all Trichinella infections that occurred in animals and humans were attributed to *T. spiralis*. Nowadays, eight species and four genotypes within two clades (encapsulated and non-encapsulated) are recognized in this genus [4,180]. Trichinellosis is caused by *Trichinella* larvae that are encysted in muscle tissue of domestic or wild animal meat. The domestic pig is considered as the most important source of human infection worldwide. However, in the past few decades, wild boar and horse meat have played similar role [181]. Infection is characterized by fever, diarrhea, periorbital oedema, and myalgia. Many severe complications like myocarditis, thromboembolic disease, and encephalitis may occur [15,181]. Europe has issued official regulations, which provide for the control of Trichinella in meat to improve consumers safety [9].

Therefore, *Trichinella* spp. is not just a hazard to public health, but also an economic problem in porcine animal production. Due that many scientific groups are working on methods to control and elimination of this parasite from the food chain. Proteomics methods are also used to help solve this problem.

Liu et al., using iTRAQ method, has described differentially regulated proteins in the three stages of *T. spiralis*—adult (Ad), muscle larvae (ML), and newborn larvae (NBL) [182]. A total of 4691 proteins were identified in all the stages, of which 1067 were differentially regulated. Different work performed on *T. spiralis* used label-free LC–MS/MS to determine the proteome differences between *T. spiralis* ML and intestinal infective larvae at the molting stage [183]. A total of 2885 proteins were identified, of which 323 were differentially regulated. These proteins were involved in regulation of cuticle synthesis, remodeling and degradation, and hormonal regulation of molting. In another study conducted on *T. britovi* (the second most common species), somatic extracts obtained from ML and Ad were separated using 2-DE coupled with immunoblot analysis. Then, the protein spots were identified by LC-MS/MS [184]. A total of 272 proteins were identified in the proteome of *T. britovi* Ad, and 261 in ML. Somatic cell extracts of Ad and ML were specifically recognized by *T. britovi*-infected swine serum 10 days after infection, with a total of 70 prominent proteins [184]. Proteomic analyses of species specific antigens were also performed with the use of MALDI-TOF and MALDI-TOF/TOF [185,186]. Potentially immunogenic proteins of the encapsulated (*T. spiralis*) and non-encapsulated (*T. pseudospiralis*, *T. papuae*) species were also investigated [187], and such proteins were identified by LC-MS/MS. Then, their possible functions were determined using gene ontology analysis. Host–parasite interactions were also analyzed by investigation of surface and excretory-secretory proteins of *Trichinella* spp. The surface proteins of *T. spiralis* muscle larvae were detected by 2-DE and MS. The 2-DE analysis detected about 33 protein spots, of which 14 were identified in the serum of mice infected with *T. spiralis*, and 12 were successfully identified by MALDI-TOF/TOF-MS [188]. The same group, using shotgun LC-MS/MS, performed comparative proteomic analysis and described surface protein profiles of ML and intestinal infective larvae [189]. A total of 41 proteins were shared by both stages, while ML had 85 and intestinal infectious larvae had 113 stage-specific proteins. Certain proteins (for example, putative onchocystatin) were involved in host-parasite interactions. Excretory–secretory proteins, as the most important products of host–parasite interaction, were investigated in the latest studies on *T. spiralis*, *T. pseudospiralis* and *T. britovi* [190–192].

2.3.3. Taenia spp.

The terms "cysticercosis" and "taeniosis" respectively refer to foodborne zoonotic infections with larval and adult tapeworms of the genus *Taenia*. The larvae of these tapeworms are meat-borne (beef or pork) and the adult stage is an obligate parasite of the human intestine [193]. *T. solium* (pork) and *T. saginata* (beef) are the most important causes of taeniosis in Europe [36]. Within the European Union, certain countries can acquire *T. solium* infection locally. There have been reports of pig infections in Hungary, Lithuania, Austria, Estonia, Romania, and Poland [194], while there were only sporadic imported cases in other countries. Humans obtain tapeworms by eating raw or undercooked infected meat. Among these tapeworms, *T. solium* is exclusive because the cysticercus stage can also infect humans directly. Human cysticercosis is acquired by accidental ingestion of *T. solium* cysticerci excreted in host feces. In humans, cysticerci may lodge in the brain and cause neurocysticercosis [195]. Taeniasis in humans is of minor clinical significance; usually, asymptomatic or symptoms are mild and non-specific (abdominal pain, weight loss, nausea, diarrhea or constipation and itching caused by proglottids, which might be passed through the anus) [15,193]. Nevertheless, cysticercosis does have major clinical significance. Intracranial hypertension and epilepsy are the most common clinical manifestations [194].

So far, to our knowledge, proteomics has described the fallowing main *Taenia* spp. features: total protein composition of cysticerci of *T. solium* by 2-DE [196]; in *T. solium*, using LC-MS/MS, a set of oncosphere proteins involved in gut penetration and immune evasion machineries in adhesion [197]; candidate antigens through immunoproteomics [198–201]; *T. solium* cysts proteomes obtained from different host tissues [202,203]; saline vesicular extract proteins of *T. solium* [204]; and *T. solium* excretory-secretory proteome [205].

2.3.4. Sarcocystis spp.

In pigs, three species of *Sarcocystis* were found: *S. miescheriana*, *S. porcifelis*, and *S. suihominis*. However, only *S. suihominis* can cause human infections when eating raw pork [206]. *S. suihominis* has an obligatory two-host life cycle. Sporocysts are shed in the feces of humans or chimpanzees, rhesus and cynomolgus monkeys (definitive host), and pigs (the intermediate host). In pigs, parasites are encapsulated in muscle tissue, but usually do not cause pathological changes or symptoms [193,207]. Infection can be asymptomatic or symptomatic (nausea, loss of appetite, stomach pain, vomiting, diarrhea, difficulty in breathing, and rapid pulse) [193]. Sarcosporidiosis is a self-limiting infection and treatment is not known.

Until now, to our knowledge, there is no published report at the proteomic analysis of *S. suihominis*.

2.4. Seafood-Borne Parasitic Species

Fish meat can be infected by a variety of parasites, which can cause human infections when eaten raw or undercooked. Additionally, various species of shellfish (mollusks and crustaceans) can be consumed by people when infected by different stages of many parasites. In addition, in many parts of the world, the term "seafood" has been extended to freshwater organisms consumed by humans, so all edible aquatic organisms can be called "seafood", including aquatic plants. Due that, in this work and according to the ranking prioritizing foodborne parasites in Europe, we describe not only sea-species parasites but also freshwater-parasitic species. Among seafood-borne parasitic infection, according the European ranking [36], the most important parasites are Anisakidae (seventh/25), Opisthorchiidae (11th/25), *Angiostrongylus cantonesis* (13th/25), *Diphyllobothrium* spp. (16th/25), *Paragonimus* spp. (23rd/25), and Heterophyidae (24th/25). Due to emerging number of Anisakidae infections in Europe and strong allergic reaction to, e.g., *Anisakis simplex* s.s., we decided to discuss the Anisakidae family in a separate section.

2.4.1. Opisthorchiidae

Opisthorchiasis is a trematode infection caused by species of the family Opisthorchiidae, specifically, *Opisthorchis viverrini* and *O. felineus* [30]. It is calculated that around 10 million people have been infected with *O. viverrini* [208], and 67 million are at risk of infection [209]. The freshwater snail is a first intermediate host of *O. viverrini*, while the second intermediate hosts include several freshwater cyprinid fish species [210]. Freshwater fish dishes infected with metacercariae have are the main source of infection of this parasites to humans [211]. Human opisthorchiasis is typically asymptomatic and therefore results in chronic inflammatory disease; this chronic inflammation can develop into the cholangiocarcinoma [212]. Thus, *O. viverrini* has been classified by the International Agency for Research on Cancer as a group 1 carcinogen. The only way to reduce the percent of cholangiocarcinoma cases, where the causative agent was *O. viverrini*, is to reduce the prevalence of opisthorchiasis through the use of praziquantel—an anthelminthic drug [212]. Unfortunately, this drug is at risk of resistance, and studies performed on *O. viverrini* could help develop efficient methods to reduce the prevalence of opisthorchiasis and induced by *O. viverrini* cholangiocarcinoma [212].

Comparative 2-DE analysis was used to highlight proteins that are significantly modulated during the maturation stage of *O. viverrini*. The differentially regulated proteins in the juvenile/adult form of the parasite are thought to be important for survival and pathogenesis. Compared with the one-week-old juvenile fluke, 35 protein spots in four-week-old adults were differentially regulated. [213]. Moreover, using proteomics (QTRAP MS/MS) Mulvenna et al. [214] characterized 300 proteins from the *O. viverrini* excretory-secretory products. In addition, more than 160 tegumental proteins were identified using sequential solubilization of isolated teguments, and some of them were located on the surface membrane of the tegument by localizing with fluorescence microscopy. The several proteins functions are still unknown [214]. Studies on proteomes of intermediate hosts of *O. viverrini* were also conducted. Proteomic profile using iTRAQ labelling technology of *Bithynia siamensis goniomphalos* snails upon infection with the *O. viverrini* was characterized [215]. This study indicates that motor proteins, and stress-related proteins are greatly upregulated after infection. In addition, the expression level of peroxiredoxins was reduced in infected *Bithynia*. Using sequential window acquisition of all theoretical spectra mass spectrometry (SWATH-MS), the protein composition of the hemolymph of *B. siamensis goniomphalos* infected with *O. viverrini* was described. The analysis revealed the presence of 242 and 362 proteins in the plasma and hemocytes, respectively [216]. Among them, the 117 and 145 proteins showed significant differences after opisthorchiasis in plasma and hemocytes, respectively. Suwannatrai et al. [216], among proteins with significantly different expression, found proteins strongly associated with immune response and proteins belonging to the structural and motor categories.

Although there are still few proteomics studies on *O. viverrini* and its hosts, many of the discovered proteins have become potential candidates for diagnostic biomarkers or new drug development.

2.4.2. Angiostrongylus cantonesis

Angiostrongylus cantonensis (*A. cantonensis*) is a parasitic nematode that occasionally causes angiostrongyliasis in humans. Its main clinical manifestation is eosinophilic meningitis [217]. Human infections are acquired by ingestion of raw or undercooked snails or slugs, paratenic hosts such as prawns, or contaminated vegetables that contain the infective larvae. After swallowing, the infective larvae are digested from these carriers and invade the intestinal tissues, causing human enteritis, and then pass through the liver. When the worm moves through the lungs, cough, rhinorrhea, sore throat, discomfort and fever occur. In about 14 days, the larvae reach the central nervous system, followed by eosinophilic meningitis and eosinophilia [217,218]. In many patients, the larvae can also move to the eyes and cause ocular angiostrongyliasis, accompanied by visual disturbances, such as diplopia or strabismus [219,220]. Detection of *A. cantonensis* in cerebrospinal fluid or the ocular chamber confirms the disease in humans. However, the percentage of confirmed cases is very low. The history of eating intermediate or paratenic hosts in medical interview is essential for the diagnosis [217]. The combination of corticosteroids and anthelmintics has been commonly used to treat this disease [217].

A. cantonensis has been widely studied using proteomic methods. The protein expression profiles of the parasite's infective third and pathogenic five stage larvae were compared by proteomics technology [221]. Isolated protein samples were separated by 2-DE, and analyzed by MALDI-TOF MS. Of the 100 protein spots identified, 33 were from L3, while 67 from L5 and 63 had known identities, and 37 were hypothetical proteins. There were 15 spots of stress proteins, and heat shock protein 60 was the most frequently found stress proteins in L5. Moreover, four protein spots were identified in the serum of the rat host by Western blotting. These changes may reflect the development of L3 from the poikilothermic snails to L5 in the homoeothermic rats [221]. The proteomes of different life stages of *A. cantonensis* were studied more widely by Huang et al. [222], who extracted soluble proteins from various stages of the *A. cantonensis* life cycle (female adults, male adults, the fifth-stage female larvae, the fifth-stage male larvae, and third-stage larvae), separated those proteins using 2D-DIGE and analyzed the gel images. Proteomics analysis yielded a total of 183 different protein spots. Through MALDI-TOF MS/MS, 37 proteins were found with a high confidence score (around 95%). Among them, 29 proteins were identified as cytoskeleton-related proteins and functional proteins [222]. The latest study aimed to identify and characterize the excretory-secretory protein profile of *A. cantonensis* adult larvae [223]. A total of 51 spots were identified using 2-DE. Then, approximately 254 proteins were identified by LC-MS/MS and further classified according to their biological functions. Finally, in the pool of excretory-secretory products of *A. cantonensis* the immunoreactive proteins were identified, including proteins like, disulphide isomerase, putative aspartic protease or annexin [223].

All this information may be useful for discovering biomarkers to diagnose highly dangerous angiostrongyliasis.

2.4.3. *Diphyllobothrium* spp.

Diphyllobothriasis is caused by flatworms of the genus *Diphyllobothrium*, and is acquired by ingestion of larval stages (plerocercoids) present in raw or undercooked fish [224]. *D. latum* is the main species infecting humans. Worms usually reside in the ileum and rarely attach to the bile ducts. In Switzerland, Italy, and France around lakes, reports of diphyllobothriasis have increased, where raw or undercooked perch was consumed. In some countries, previously considered disease-free (Austria, Czech Republic, Belgium, Netherlands and Spain), few cases have been reported, probably related to the consumption of imported raw fish [225]. Although most *Diphyllobothrium* species are large (2–15 m) and can have a mechanical effect on the host, infections are often asymptomatic. About 20% of people experience diarrhea, discomfort and abdominal pain. Other symptoms may also occur, such as fatigue, constipation, pernicious anemia, headache, and allergic reactions. Although large-scale infection is not common, it may cause intestinal obstruction, and the migrating segments can cause cholecystitis or cholangitis [224]. A single dose of praziquantel is highly effective against diphyllobothriasis [225].

Despite the high prevalence of this disease (about 20 million people infected worldwide), to our knowledge, *D. latum* proteome has not been described.

2.4.4. *Paragonimus* spp.

There are about 15 species of *Paragonimus* known to infect humans, while *P. westermani* is the most common etiological agent of human paragonimiasis in Europe [226]. After ingesting raw or undercooked freshwater crustaceans (such as crabs, shrimps or crayfish), humans or other final hosts (carnivores) can become infected. The metacercariae excyst in the small intestine and passes through the intestinal wall into the abdominal cavity before it migrates through the sub-peritoneal tissues, and finally enters the lung where maturation occurs. Eggs of adult individuals, which are coughed up and ejected by spitting with the sputum or swallowed and passed in the feces, hatch, and miracidia invade freshwater snails. Then the cercariae emerge, and crustacea consuming may get infected by consuming it directly or eating infected snails containing the fully developed cercariae [30]. The presence of paragonimiasis can cause bleeding, inflammation, lung parenchymal necrosis and fibrotic cysts. In lung paragonimiasis, the most obvious symptom is chronic cough, accompanied by brown and bloody pneumonia-like sputum [30,226].

Despite the importance of the disease, little information about the proteomics of *Paragonimus* spp. can be found. The only analyzed excretory–secretory products of adult *P. westermani* using 2-DE coupled to MS [227]. In this study 25 different proteins were identified, some of which are highly representative, such as cysteine proteases. In addition, three previously unknown cysteine proteases were also identified by MALDI-TOF/TOF MS, and most of them are reactive to serum from patients with paragonimiasis. Park et al. [228] suggested that a new drug for paragonimiasis could be designed, focusing on exploring inhibitors of cysteine proteases.

2.4.5. Heterophyidae

In humans, heterophyidiasis and metagonimiasis is associated mainly with species of *Heterophyes* or *Metagonimus*, respectively. Those diseases are the best-known associated with heterophyid parasitism [229]. Humans can get infected usually by eating raw, undercooked or under-processed fish [230]. The two most widespread species of heterophyids are *H. heterophyes* and *M. yokogawai*. There are evidences in European countries of infections caused by *H. heterophyes* (Spain, Italy, Greece, Turkey) and *M. yokoagwai* (Bulgaria, the Czech Republic, Romania, Serbia, Spain, and Ukraine). Intriguing, there have been no reports of *M. yokoagwai* infecting humans [229].

The parasitic hosts of *H. heterophyes* include dogs, cats, pigs, fish-eating birds, and other fish-eating mammals. Adult worms live in the small intestine of vertebrate hosts and the gastropods from the genus *Semisulcospira* are the earliest intermediate hosts, where cyprinid fish are mainly a second intermediate host [230]. Most heterophyids infections have no clinical consequences, but severe ones are causing gastro-intestinal problems [230].

The genome of *M. yokogawai* has not been sequenced. To our knowledge, there is no information about proteomics of *M. yokogawai*, and of *H. heterophyes*.

3. Targeted Approach—Proteomics Methods Proposed to Use for Detection of Selected FBPs in Food

The one of the major worldwide concern is food safety [231]. Foods contaminated by a range of FBPs is a serious issue causing economic losses in the food sector because it undermines consumer confidence and lowers the demand for potentially infected food products [7,15]. There is a need for technologies that can detect pathogens quickly and early to ensure enhanced food safety. To date, no guidelines or microbiological criteria exist for most FBPs in food products. Advanced proteomic-based methods, like MS, have a great potential for FBPs identification in food. However, detection methods currently available for the selected FBPs are mostly established on standard parasitological approaches (e.g., FBP detection in food product sample by visual examination or microscopy) or on by the polymerase chain reaction (PCR) and enzyme-linked immunosorbent assay (ELISA) [15,55,232,233]. Moreover, most of the conventional proteomic techniques such as ion exchange chromatography, size exclusion chromatography, affinity chromatography, ELISA, western blotting SDS-PAGE, 2-DE, and 2DE-DIGE are used for detection of fungi and bacteria [234–236]. MALDI-TOF MS, surface-enhanced laser desorption ionization time of flight mass spectrometry (SELDI-TOF MS), LC-MS/MS, isotope-coded affinity tags (ICAT), and iTRAQ are the central among current proteomics. The discovery proteomic workflow used for identification of biomarkers of parasitic infections is essential for diagnostic purposes and new treatment inventions. Biomarkers can be either proteins of the parasite itself or host proteins responding to infection. Then, the targeted proteomics could be used to search with high precision, sensitivity and reproducibility the peptide biomarkers selected in the discovery phase in patient biological fluids or in food products [17,237]. Although identification of protein targets has been done in many FBPs, the proteomic methods to detect them are still rare.

SELDI is one of the most used in the studies published about parasitic diseases. This proteomic method has been applied to investigate the serum biomarkers of African trypanosomiasis [238], fascioliasis [239], cysticercosis [240], and Chagas disease [241]. These studies have focused on identifying a "proteomic fingerprint" in infected people's/animals' serum—a unique configuration of parasite proteins that indicate a specific pathophysiological state. Even so, these methods are not widely used. In our opinion, there is a colossal need to diagnose foodborne infections using new methodologies, such as MS and biomarkers detection, according to their extreme specificity and possible increase of the detection rate [242].

Another factor that has attracted increasing attention to FBPs is the increasing demand for protein-rich foods such as fish. [243]. Thus, seafood-borne parasitic infections, such as those caused by Heterophyidae, Opisthorchiidae, and Anisakidae, are emerging ones, and development of advanced and more accurate methods for identification, monitoring, and assessing of FBPs during production, processing, and storage should be now worldwide concern.

Therefore, in this part, we focused on the brief description of the proteomic methods proposed to use for Anisakidae detection in food products.

Anisakidae

The consumption of raw or unprocessed fish infected with cosmopolitan nematodes belonging to the Anisakidae family may lead to anisakidosis [244]. Known human-infecting Anisakids species include members of the *Anisakis simplex* complex (*A. simplex* sensu stricto, *A. pegreffii*, *A. berlandi*), the *Pseudoterranova decipiens* complex (*P. decipiens* sensu stricto, *P. azarasi*, *P. cattani*, and others), and the *Contracecum osculatum* [244]. Among them, *A. simplex* is the most commonly involved in human infections, and the disease caused by the *Anisakis* genus is called anisakiasis [245,246]. The life cycle of *A. simplex* is complex and involves four larval stages parasitizing several intermediate and paratenic hosts (fish, cephalopods, and crustaceans) and the adult stage parasitizing marine mammals (seals, dolphins, and whales). Humans can be accidentally infected by eating raw or undercooked fish or seafood contaminated by third-stage (infective) development stage [244,247].

The ingestion of viable larvae might lead to gastrointestinal symptoms (abdominal pain, nausea, vomiting, diarrhea), which may be associated with mild to severe allergic reactions, and the clinical symptoms most often are such as rhinitis, urticaria, and, in worst cases, anaphylactic shock [245,246]. Although, cooking (or freezing) is expected to kill the parasites, it might not decrease its allergenicity, because *A. simplex* allergens have high heat and frost resistance; and sensitization may occur after consumption [247].

Proteomic studies on *A. simplex* were reviewed meticulously and accurately by D'Amelio et al. [248]. In brief, using MALDI-TOF/TOF analysis, differentially expressed proteins for *A. simplex* s.s., *A. pegreffii*, and their hybrid were described, and potential new allergens were identified. In a similar study, Fæste et al. [249] characterized, using sera from *A. simplex*-sensitized patients, potential allergens. Additionally, biomarker (peptides) for relevant *A. simplex* proteins were described. Most of all, *A. simplex* allergens have been identified and characterized [248]. Recently, for the first time, the global proteome of the third and fourth stage larvae of *A. simplex* was analyzed using quantitative proteomics based on tandem mass tag (TMT) [31]. In addition, the response to the invasive larvae of *A. simplex* s.s. to ivermectin (anthelminthic drug) was also evaluated using TMT-based methodology [250].

The discovery of potential allergens has prompted scientists to use a targeted approach and develop methodologies for detection of Anisakids in food products. Lately, the method for detection of *A. simplex* allergens in fresh, frozen, and cooked fish meat was proposed with the use of immunoglobulin G (IgG) immunoblotting [251]. Recently, many Anisakids proteins have been identified through LC-MS/MS-based proteomics, which laid the foundation for the rise of detection methods of *A. simplex* in fish. Fæste et al. [252] have shown by ELISA, immunostaining, and MS, proteins of *A. simplex* in salmon meant for use in sushi and other fish products on the Norwegian market. The same group proposed two more methodologies for *A. simplex* protein detection in fish [253]. Both were based on multiple reaction monitoring (MRM)-MS/MS, which is applied to quantify previously identified target peptides by measuring specific precursor-to-product ion transitions. Both proposed methodologies, the label-free semi-quantitative nLC-nESI-Orbitrap-MS/MS and the heavy peptide-applying absolute-quantitative (AQUA) LC-TripleQ-MS/MS use unique reporter peptides derived from Anisakids hemoglobin and SXP/RAL-2 protein as analytes.

Recently, the analysis performed by Carrera et al. [254] showed possible practical use of peptide biomarkers in food industry. The discovery phase was based on the isolation of heat-stable proteins of *A. simplex*, *P. krabbei*, and *P. decipiens*, then the use of accelerated in-solution trypsin digestions under an ultrasonic field provided by high-intensity focused ultrasound (HIFU) and the monitoring of several peptide biomarkers by parallel reaction monitoring (PRM) mass spectrometry in a linear ion trap mass spectrometer. The target detection step showed the same proportional relationships between the proposed peptide biomarkers that spiked in hake protein extracts, like those of the buffer diluted sample, which confirms the effectiveness of the PRM method in real fish samples. This method can quickly detect Anisakids in less than 2 h [254], and if it is made a part of a control protocol defined by food safety authorities, it may facilitate testing and thus increase consumer safety.

4. Conclusions and Future Directions

These days the Proteomics provides a great tool in both basic and applied parasitology. Proteomics methods, like SDS-PAGE, 2-DE, and LC-MS/MS combined with bioinformatic tools, have become common in modern helminth parasitology research. These methods have the potential to identify differentially regulated proteins in diverse parasite development stages or in response to drugs, as well as to describe the composition of parasitic extracellular vesicles. The discovery of new parasite proteins, consequently, might help to find candidates for the parasites' detection, modern therapies, and vaccines. Although proteomics has broadened the parasitologists view on the parasites' physiology and parasite–host interactions and crosstalk, little is known about many foodborne parasites and diseases caused by them, including heterophyiasis (*H. heterophyes*), diphyllobothriasis (*D. latum*), sarcosporidiosis (*Sarcocystis* spp.), or paragonimiasis (*P. westermani*). Despite this, great attempts have been made to point out and highlight the importance and benefits of using advanced proteomics methods for the detection of FBPs in food.

Nevertheless, the use of proteomic tools, including software for equipment, databases, and the requirement of skilled personnel, significantly increases costs and therefore limits their wider use. Currently, the potential of proteomics has been used as a finding tool for novel biomarkers, which can then be integrated into uncomplicated diagnostic methods based on inexpensive technologies such as antigen detection in immunochromatographic analysis and other biosensors. Prospective strategies should concentrate on developing proteomic methodologies that are accessible to analytical laboratories or even to the laboratories of food processing plants, with reduced expenses and time-to-result. Furthermore, developed methods should be appropriate for a broad spectrum of different food types and parasite development stages contaminating them, because single methods or biomarkers might not be suitable for the detection of different FBPs stages in a variety of types of food.

Parasite proteins discovery might help developing methodologies for the rapid detection of these contaminants in food products and fill the gap in fields of animal production, agriculture, food processing, and storage, thus benefiting human health. We hope that proteomic methods will be key in opening the door into the increases of detection rate of foodborne parasites in foods, as well as into the reduction of the prevalence of FBPs caused diseases.

Author Contributions: R.S. wrote the initial version of the manuscript. E.Ł.-B. and M.C. assisted in the completion and reviewing of the final version of the manuscript. All authors have read and agreed to the published version of the manuscript.

Funding: This work was funded by the "Development Program of the University of Warmia and Mazury in Olsztyn," co-financed by the European Union under the European Social Fund from the Operational Program Knowledge Education Development. Robert Stryiński is a recipient of a scholarship from the program

"Interdisciplinary Doctoral Studies in Biology and Biotechnology" (project number POWR.03.05.00-00-Z310/17), which is funded by the European Social Fund. This work was additionally co-funded by the GAIN-Xunta de Galicia (project number IN607D 2017/01) and the Spanish AEI/EU-FEDER PID2019-103845RB-C21 project. Mónica Carrera is supported by the Ramón y Cajal contract (Ministry of Science and Innovation of Spain).

Conflicts of Interest: The authors declare no conflict of interest. The funders had no role in the design of the study; in the collection, analyses, or interpretation of data; in the writing of the manuscript, or in the decision to publish the results.

References

1. Trevisan, C.; Torgerson, P.R.; Robertson, L.J. Foodborne Parasites in Europe: Present Status and Future Trends. *Trends Parasitol.* **2019**, *35*, 695–703. [CrossRef] [PubMed]
2. Pozio, E. How globalization and climate change could affect foodborne parasites. *Exp. Parasitol.* **2020**, *208*, 107807. [CrossRef] [PubMed]
3. Dorny, P.; Praet, N.; Deckers, N.; Gabriel, S. Emerging food-borne parasites. *Vet. Parasitol.* **2009**, *163*, 196–206. [CrossRef] [PubMed]
4. Murrell, K.D. Zoonotic foodborne parasites and their surveillance. *OIE Rev. Sci. Tech.* **2013**, *32*, 559–569. [CrossRef]
5. EFSA and ECDC The European Union summary report on trends and sources of zoonoses, zoonotic agents and food-borne outbreaks in 2017. *EFSA J.* **2018**, *16*. [CrossRef]
6. Poulin, R. Parasite biodiversity revisited: Frontiers and constraints. *Int. J. Parasitol.* **2014**, *44*, 581–589. [CrossRef] [PubMed]
7. Koutsoumanis, K.; Allende, A.; Alvarez-Ordóñez, A.; Bolton, D.; Bover-Cid, S.; Chemaly, M.; Davies, R.; De Cesare, A.; Herman, L.; Hilbert, F.; et al. Public health risks associated with food-borne parasites. *EFSA J.* **2018**, *16*. [CrossRef]
8. Crompton, D.W.T. How Much Human Helminthiasis Is There in the World? *J. Parasitol.* **1999**, *85*, 397–403. [CrossRef]
9. European Commission Implementing Regulation (EU) 2015/1375 of 10 August 2015 laying down specific rules on official controls for Trichinella in meat. *Off. J. Eur. Union* **2015**, *L212*, 7–34.
10. European Commission Regulation (EC) No 854/2004 of the European Parliament and of the Council of 29 April 2004 laying down specific rules for the organisation of official controls on products of animal origin intended for human consumption. *Off. J. Eur. Union* **2004**, *2003*, 83.
11. European Commission Regulation (EC) No 853/2004 of the European Parlament and of the Council of 29 April 2004 laying down specific hygiene rules for on the hygiene of foodstuffs. *Off. J. Eur. Union* **2004**, *L 139*, 55.
12. Åberg, R.; Sjöman, M.; Hemminki, K.; Pirnes, A.; Räsänen, S.; Kalanti, A.; Pohjanvirta, T.; Caccio, S.M.; Pihlajasaari, A.; Toikkanen, S.; et al. Cryptosporidium parvum Caused a Large Outbreak Linked to Frisée Salad in Finland, 2012. *Zoonoses Public Health* **2015**, *62*, 618–624. [CrossRef] [PubMed]
13. McKerr, C.; Adak, G.K.; Nichols, G.; Gorton, R.; Chalmers, R.M.; Kafatos, G.; Cosford, P.; Charlett, A.; Reacher, M.; Pollock, K.G.; et al. An outbreak of cryptosporidium parvum across England and Scotland associated with consumption of fresh pre-cut salad leaves, May 2012. *PLoS ONE* **2015**, *10*, e0125955. [CrossRef] [PubMed]
14. McKerr, C.; O'Brien, S.J.; Chalmers, R.M.; Vivancos, R.; Christley, R.M. Exposures associated with infection with Cryptosporidium in industrialised countries: A systematic review protocol. *Syst. Rev.* **2018**, *7*, 70. [CrossRef] [PubMed]
15. Chalmers, R.M.; Robertson, L.J.; Dorny, P.; Jordan, S.; Kärssin, A.; Katzer, F.; La Carbona, S.; Lalle, M.; Lassen, B.; Mladineo, I.; et al. Parasite detection in food: Current status and future needs for validation. *Trends Food Sci. Technol.* **2020**, *99*, 337–350. [CrossRef]
16. Cifuentes, A. Food Analysis and Foodomics. *J Chromatogr. A.* **2009**, *1216*, 7109. [CrossRef]
17. Gallardo, J.M.; Ortea, I.; Carrera, M. Proteomics and its applications for food authentication and food-technology research. *Trends Anal. Chem.* **2013**, *52*, 135–141. [CrossRef]
18. Herrero, M.; Simó, C.; García-Cañas, V.; Ibáñez, E.; Cifuentes, A. Foodomics: MS-based strategies in modern food science and nutrition. *Mass Spectrom. Rev.* **2012**, *31*, 49–69. [CrossRef]
19. Siciliano, R.A.; Uzzau, S.; Mazzeo, M.F. Editorial: Proteomics for studying foodborne microorganisms and their impact on food quality and human health. *Front. Nutr.* **2019**, *6*, 104. [CrossRef]
20. Pandey, A.; Mann, M. Proteomics to study genes and genomes. *Nature* **2000**, *405*, 837–846. [CrossRef]

21. Ong, S.E.; Mann, M. Mass Spectrometry–Based Proteomics Turns Quantitative. *Nat. Chem. Biol.* **2005**, *1*, 252–262. [CrossRef] [PubMed]
22. Emery-Corbin, S.J.; Grüttner, J.; Svärd, S. Transcriptomic and proteomic analyses of Giardia intestinalis: Intestinal epithelial cell interactions. In *Advances in Parasitology*; Academic Press: London, UK, 2020; Volume 107, pp. 139–171. ISBN 9780128204757.
23. Walhout, M.; Vidal, M.; Dekker, J. *Handbook of Systems Biology*; Elsevier: Amsterdam, The Netherlands, 2013; ISBN 9780123859440.
24. 50 Helminth Genomes Initiative. Available online: http://www.sanger.ac.uk/resources/downloads/helminths/ (accessed on 18 July 2020).
25. Ginger, M.L.; McKean, P.G.; Burchmore, R.; Grant, K.M. Proteomic insights into parasite biology. *Parasitology* **2012**, *139*, 1101–1102. [CrossRef] [PubMed]
26. Marcilla, A.; Trelis, M.; Cortés, A.; Sotillo, J.; Cantalapiedra, F.; Minguez, M.T.; Valero, M.L.; Sánchez del Pino, M.M.; Muñoz-Antoli, C.; Toledo, R.; et al. Extracellular Vesicles from Parasitic Helminths Contain Specific Excretory/Secretory Proteins and Are Internalized in Intestinal Host Cells. *PLoS ONE* **2012**, *7*, e45974. [CrossRef] [PubMed]
27. Coakley, G.; Maizels, R.M.; Buck, A.H. Exosomes and Other Extracellular Vesicles: The New Communicators in Parasite Infections. *Trends Parasitol.* **2015**, *31*, 477–489. [CrossRef] [PubMed]
28. Riaz, F.; Cheng, G. Exosome-like vesicles of helminths: Implication of pathogenesis and vaccine development. *Ann. Transl. Med.* **2017**, *5*, 10–12. [CrossRef] [PubMed]
29. Młocicki, D.; Sulima, A.; Bień, J.; Näreaho, A.; Zawistowska-Deniziak, A.; Basałaj, K.; Sałamatin, R.; Conn, D.B.; Savijoki, K. Immunoproteomics and Surfaceomics of the Adult Tapeworm Hymenolepis diminuta. *Front. Immunol.* **2018**, *9*, 2487. [CrossRef] [PubMed]
30. Toledo, R.; Bernal, M.D.; Marcilla, A. Proteomics of foodborne trematodes. *J. Proteom.* **2011**, *74*, 1485–1503. [CrossRef]
31. Stryiński, R.; Mateos, J.; Pascual, S.; González, Á.F.; Gallardo, J.M.; Łopieńska-Biernat, E.; Medina, I.; Carrera, M. Proteome profiling of L3 and L4 Anisakis simplex development stages by TMT-based quantitative proteomics. *J. Proteom.* **2019**, *201*, 1–11. [CrossRef]
32. Prester, L. Seafood Allergy, Toxicity, and Intolerance: A Review. *J. Am. Coll. Nutr.* **2016**, *35*, 271–283. [CrossRef]
33. Marzano, V.; Tilocca, B.; Fiocchi, A.G.; Vernocchi, P.; Levi Mortera, S.; Urbani, A.; Roncada, P.; Putignani, L. Perusal of food allergens analysis by mass spectrometry-based proteomics. *J. Proteom.* **2020**, *215*, 103636. [CrossRef]
34. Carrera, M.; Piñeiro, C.; Martinez, I. Proteomic Strategies to Evaluate the Impact of Farming Conditions on Food Quality and Safety in Aquaculture Products. *Foods* **2020**, *9*, 1050. [CrossRef] [PubMed]
35. Aebersold, R.; Bensimon, A.; Collins, B.C.; Ludwig, C.; Sabido, E. Applications and Developments in Targeted Proteomics: From SRM to DIA/SWATH. *Proteomics* **2016**, *16*, 2065–2067. [CrossRef] [PubMed]
36. Bouwknegt, M.; Devleesschauwer, B.; Graham, H.; Robertson, L.J.; van der Giessen, J.W. Prioritisation of food-borne parasites in Europe, 2016. *Eurosurveillance* **2018**, *23*. [CrossRef] [PubMed]
37. Ryan, U.; Hijjawi, N.; Xiao, L. Foodborne cryptosporidiosis. *Int. J. Parasitol.* **2018**, *48*, 1–12. [CrossRef]
38. Valenzuela, O.; González-Díaz, M.; Garibay-Escobar, A.; Burgara-Estrella, A.; Cano, M.; Durazo, M.; Bernal, R.M.; Hernandez, J.; Xiao, L. Molecular Characterization of *Cryptosporidium* spp. in Children from Mexico. *PLoS ONE* **2014**, *9*, e96128. [CrossRef]
39. Chalmers, R.M.; Davies, A.P. Minireview: Clinical cryptosporidiosis. *Exp. Parasitol.* **2010**, *124*, 138–146. [CrossRef]
40. Ryan, U.; Zahedi, A.; Paparini, A. Cryptosporidium in humans and animals—A one health approach to prophylaxis. *Parasite Immunol.* **2016**, *38*, 535–547. [CrossRef]
41. Bones, A.J.; Jossé, L.; More, C.; Miller, C.N.; Michaelis, M.; Tsaousis, A.D. Past and future trends of Cryptosporidium in vitro research. *Exp. Parasitol.* **2019**, *196*, 28–37. [CrossRef]
42. Sanderson, S.J.; Xia, D.; Prieto, H.; Yates, J.; Heiges, M.; Kissinger, J.C.; Bromley, E.; Lal, K.; Sinden, R.E.; Tomley, F.; et al. Determining the protein repertoire of *Cryptosporidium parvum* sporozoites. *Proteomics* **2008**, *8*, 1398–1414. [CrossRef]
43. Siddiki, A.M.A.M.Z.; Wastling, J.M. Charting the proteome of Cryptosporidium parvum sporozoites using sequence similarity-based BLAST searching. *J. Vet. Sci.* **2009**, *10*, 203–210. [CrossRef]

44. Snelling, W.J.; Lin, Q.; Moore, J.E.; Millar, B.C.; Tosini, F.; Pozio, E.; Dooley, J.S.G.; Lowery, C.J. Proteomics analysis and protein expression during sporozoite excystation of Cryptosporidium parvum (coccidia, apicomplexa). *Mol. Cell. Proteom.* **2007**, *6*, 346–355. [CrossRef]
45. Shrivastava, A.K.; Kumar, S.; Sahu, P.S.; Mahapatra, R.K. In silico identification and validation of a novel hypothetical protein in *Cryptosporidium hominis* and virtual screening of inhibitors as therapeutics. *Parasitol. Res.* **2017**, *116*, 1533–1544. [CrossRef] [PubMed]
46. Ankarklev, J.; Jerlström-Hultqvist, J.; Ringqvist, E.; Troell, K.; Svärd, S.G. Behind the smile: Cell biology and disease mechanisms of Giardia species. *Nat. Rev. Microbiol.* **2010**, *8*, 413–422. [CrossRef] [PubMed]
47. Ringqvist, E.; Palm, J.E.D.; Skarin, H.; Hehl, A.B.; Weiland, M.; Davids, B.J.; Reiner, D.S.; Griffiths, W.J.; Eckmann, L.; Gillin, F.D.; et al. Release of metabolic enzymes by Giardia in response to interaction with intestinal epithelial cells. *Mol. Biochem. Parasitol.* **2008**, *159*, 85–91. [CrossRef] [PubMed]
48. Dubourg, A.; Xia, D.; Winpenny, J.P.; Al Naimi, S.; Bouzid, M.; Sexton, D.W.; Wastling, J.M.; Hunter, P.R.; Tyler, K.M. Giardia secretome highlights secreted tenascins as a key component of pathogenesis. *Gigascience* **2018**, *7*, giy003. [CrossRef]
49. Ma'ayeh, S.Y.; Liu, J.; Peirasmaki, D.; Hörnaeus, K.; Bergström Lind, S.; Grabherr, M.; Bergquist, J.; Svärd, S.G. Characterization of the Giardia intestinalis secretome during interaction with human intestinal epithelial cells: The impact on host cells. *PLoS Negl. Trop. Dis.* **2017**, *11*, e0006120. [CrossRef]
50. Wampfler, P.B.; Tosevski, V.; Nanni, P.; Spycher, C.; Hehl, A.B. Proteomics of Secretory and Endocytic Organelles in *Giardia lamblia*. *PLoS ONE* **2014**, *9*, e94089. [CrossRef]
51. Evans-Osses, I.; Mojoli, A.; Monguió-Tortajada, M.; Marcilla, A.; Aran, V.; Amorim, M.; Inal, J.; Borràs, F.E.; Ramirez, M.I. Microvesicles released from *Giardia intestinalis* disturb host-pathogen response in vitro. *Eur. J. Cell Biol.* **2017**, *96*, 131–142. [CrossRef]
52. Davids, B.J.; Liu, C.M.; Hanson, E.M.; Le, C.H.Y.; Ang, J.; Hanevik, K.; Fischer, M.; Radunovic, M.; Langeland, N.; Ferella, M.; et al. Identification of conserved candidate vaccine antigens in the surface proteome of giardia lamblia. *Infect. Immun.* **2019**, *87*, e00219-19. [CrossRef] [PubMed]
53. Müller, J.; Braga, S.; Uldry, A.C.; Heller, M.; Müller, N. Comparative proteomics of three Giardia lamblia strains: Investigation of antigenic variation in the post-genomic era. *Parasitology* **2020**. [CrossRef] [PubMed]
54. Centers for Disease Control and Prevention Cyclosporiasis. Available online: https://www.cdc.gov/parasites/cyclosporiasis/index.html (accessed on 22 July 2020).
55. Tefera, T.; Tysnes, K.R.; Utaaker, K.S.; Robertson, L.J. Parasite contamination of berries: Risk, occurrence, and approaches for mitigation. *Food Waterborne Parasitol.* **2018**, *10*, 23–38. [CrossRef] [PubMed]
56. Almeria, S.; Cinar, H.N.; Dubey, J.P. Cyclospora cayetanensis and Cyclosporiasis: An Update. *Microorganisms* **2019**, *7*, 317. [CrossRef] [PubMed]
57. Herwaldt, B.L. Cyclospora cayetanensis: A Review, Focusing on the Outbreaks of Cyclosporiasis in the 1990s. *Clin. Infect. Dis.* **2000**, *31*, 1040–1057. [CrossRef] [PubMed]
58. Prasad, K.J. Emerging and re-emerging parasitic diseases. *J. Int. Med. Sci. Acad.* **2010**, *23*, 45–50.
59. Cinar, H.N.; Qvarnstrom, Y.; Wei-Pridgeon, Y.; Li, W.; Nascimento, F.S.; Arrowood, M.J.; Murphy, H.R.; Jang, A.; Kim, E.; Kim, R.; et al. Comparative sequence analysis of Cyclospora cayetanensis apicoplast genomes originating from diverse geographical regions. *Parasit. Vectors* **2016**, *9*, 611. [CrossRef] [PubMed]
60. Liu, S.; Wang, L.; Zheng, H.; Xu, Z.; Roellig, D.M.; Li, N.; Frace, M.A.; Tang, K.; Arrowood, M.J.; Moss, D.M.; et al. Comparative genomics reveals *Cyclospora cayetanensis* possesses coccidia-like metabolism and invasion components but unique surface antigens. *BMC Genom.* **2016**, *17*, 316. [CrossRef]
61. Luna-Nácar, M.; Navarrete-Perea, J.; Moguel, B.; Bobes, R.J.; Laclette, J.P.; Carrero, J.C. Proteomic Study of Entamoeba histolytica Trophozoites, Cysts, and Cyst-Like Structures. *PLoS ONE* **2016**, *11*, e0156018. [CrossRef]
62. Mahmud, R.; Lim, Y.A.L.; Amir, A. *Medical Parasitology*; Springer International Publishing: Cham, Switzerland, 2017; ISBN 978-3-319-68794-0.
63. Leitsch, D.; Radauer, C.; Paschinger, K.; Wilson, I.B.H.; Breiteneder, H.; Scheiner, O.; Duchêne, M. *Entamoeba histolytica*: Analysis of the trophozoite proteome by two-dimensional polyacrylamide gel electrophoresis. *Exp. Parasitol.* **2005**, *110*, 191–195. [CrossRef]
64. Tolstrup, J.; Krause, E.; Tannich, E.; Bruchhaus, I. Proteomic analysis of *Entamoeba histolytica*. *Parasitology* **2007**, *134*, 289–298. [CrossRef]

65. Marion, S.; Laurent, C.; Guillén, N. Signalization and cytoskeleton activity through myosin IB during the early steps of phagocytosis in *Entamoeba histolytica*: A proteomic approach. *Cell. Microbiol.* **2005**, *7*, 1504–1518. [CrossRef]
66. Biller, L.; Matthiesen, J.; Kühne, V.; Lotter, H.; Handal, G.; Nozaki, T.; Saito-Nakano, Y.; Schümann, M.; Roeder, T.; Tannich, E.; et al. The Cell Surface Proteome of *Entamoeba histolytica*. *Mol. Cell. Proteom.* **2014**, *13*, 132–144. [CrossRef] [PubMed]
67. López-Rosas, I.; Marchat, L.A.; Olvera, B.G.; Guillen, N.; Weber, C.; Hernández de la Cruz, O.; Ruíz-García, E.; Astudillo-de la Vega, H.; López-Camarillo, C. Proteomic analysis identifies endoribouclease EhL-PSP and EhRRP41 exosome protein as novel interactors of EhCAF1 deadenylase. *J. Proteom.* **2014**, *111*, 59–73. [CrossRef] [PubMed]
68. Ujang, J.; Sani, A.A.A.; Lim, B.H.; Noordin, R.; Othman, N. Analysis of Entamoeba histolytica Membrane Proteome Using Three Extraction Methods. *Proteomics* **2018**, *18*, 1700397. [CrossRef]
69. Ujang, J.A.; Kwan, S.H.; Ismail, M.N.; Lim, B.H.; Noordin, R.; Othman, N. Proteome analysis of excretory-secretory proteins of Entamoeba histolytica HM1:IMSS via LC–ESI-MS/MS and LC–MALDI–TOF/TOF. *Clin. Proteom.* **2016**, *13*, 33. [CrossRef] [PubMed]
70. Perdomo, D.; Aït-Ammar, N.; Syan, S.; Sachse, M.; Jhingan, G.D.; Guillén, N. Cellular and proteomics analysis of the endomembrane system from the unicellular *Entamoeba histolytica*. *J. Proteom.* **2015**, *112*, 125–140. [CrossRef]
71. Leitsch, D.; Wilson, I.B.; Paschinger, K.; Duchêne, M. Comparison of the proteome profiles of *Entamoeba histolytica* and its close but non-pathogenic relative *Entamoeba dispar*. *Wien. Klin. Wochenschr.* **2006**, *118*, 37–41. [CrossRef]
72. Davis, P.H.; Zhang, X.; Guo, J.; Townsend, R.R.; Stanley, S.L. Comparative proteomic analysis of two *Entamoeba histolytica* strains with different virulence phenotypes identifies peroxiredoxin as an important component of amoebic virulence. *Mol. Microbiol.* **2006**, *61*, 1523–1532. [CrossRef]
73. Davis, P.H.; Chen, M.; Zhang, X.; Clark, C.G.; Townsend, R.R.; Stanley, S.L. Proteomic Comparison of *Entamoeba histolytica* and *Entamoeba dispar* and the Role of *E. histolytica* Alcohol Dehydrogenase 3 in Virulence. *PLoS Negl. Trop. Dis.* **2009**, *3*, e415. [CrossRef]
74. Calderaro, A.; Piergianni, M.; Buttrini, M.; Montecchini, S.; Piccolo, G.; Gorrini, C.; Rossi, S.; Chezzi, C.; Arcangeletti, M.C.; Medici, M.C.; et al. MALDI-TOF Mass Spectrometry for the Detection and Differentiation of *Entamoeba histolytica* and *Entamoeba dispar*. *PLoS ONE* **2015**, *10*, e0122448. [CrossRef]
75. Ali, I.K.M.; Haque, R.; Siddique, A.; Kabir, M.; Sherman, N.E.; Gray, S.A.; Cangelosi, G.A.; Petri, W.A. Proteomic Analysis of the Cyst Stage of *Entamoeba histolytica*. *PLoS Negl. Trop. Dis.* **2012**, *6*, e1643. [CrossRef]
76. Marquay Markiewicz, J.; Syan, S.; Hon, C.C.; Weber, C.; Faust, D.; Guillen, N. A Proteomic and Cellular Analysis of Uropods in the Pathogen *Entamoeba histolytica*. *PLoS Negl. Trop. Dis.* **2011**, *5*, e1002. [CrossRef] [PubMed]
77. Centers for Disease Control and Prevention Sparganosis. Available online: https://www.cdc.gov/dpdx/sparganosis/index.html (accessed on 28 July 2020).
78. Wiwanitkit, V. A review of human sparganosis in Thailand. *Int. J. Infect. Dis.* **2005**, *9*, 312–316. [CrossRef] [PubMed]
79. Kim, J.H.; Kim, Y.J.; Sohn, W.M.; Bae, Y.M.; Hong, S.T.; Choi, M.H. Differential protein expression in Spirometra erinacei according to its development in its final host. *Parasitol. Res.* **2009**, *105*, 1549–1556. [CrossRef]
80. Hu, D.D.; Cui, J.; Wang, L.; Liu, L.N.; Wei, T.; Wang, Z.Q. Immunoproteomic analysis of the excretory-secretory proteins from *Spirometra mansoni* sparganum. *Iran. J. Parasitol.* **2013**, *8*, 408–416.
81. Hu, D.D.; Cui, J.; Xiao, D.; Wang, L.; Liu, L.N.; Liu, R.D.; Zhang, J.Z.; Wang, Z.Q. Identification of early diagnostic antigens from *Spirometra erinaceieuropaei* sparganum soluble proteins using immunoproteomics. *Southeast Asian J. Trop. Med. Public Health* **2014**, *45*, 576–583. [PubMed]
82. Liu, W.; Tang, H.; Abuzeid, A.M.I.; Tan, L.; Wang, A.; Wan, X.; Zhang, H.; Liu, Y.; Li, G. Protein phosphorylation networks in spargana of *Spirometra erinaceieuropaei* revealed by phosphoproteomic analysis. *Parasites Vectors* **2020**, *13*, 248. [CrossRef] [PubMed]
83. Eckert, J.; Deplazes, P. Biological, Epidemiological, and Clinical Aspects of Echinococcosis, a Zoonosis of Increasing Concern. *Clin. Microbiol. Rev.* **2004**, *17*, 107–135. [CrossRef] [PubMed]
84. Moro, P.; Schantz, P.M. Echinococcosis: A review. *Int. J. Infect. Dis.* **2009**, *13*, 125–133. [CrossRef] [PubMed]

85. Moro, P.L.; Garcia, H.H.; Gonzales, A.E.; Bonilla, J.J.; Verastegui, M.; GilmanMD, R.H. Screening for cystic echinococcosis in an endemic region of Peru using portable ultrasonography and the enzyme-linked immunoelectrotransfer blot (EITB) assay. *Parasitol. Res.* **2005**, *96*, 242–246. [CrossRef]
86. WHO Informal Working Group on Echinococcosis (IWGE) Guidelines for treatment of cystic and alveolar echinococcosis in humans. *Bull. World Health Organ.* **1996**, *74*, 231–242.
87. WHO Informal Working Group on Echinococcosis (IWGE). *Puncture, Aspiration, Injection, Re-aspiration: An Option for the Treatment of Cystic Echinococcosis*; World Health Organization: Geneva, Switzerland, 2001; WHO/CDS/CSR/APH/2001.6.
88. Chemale, G.; Van Rossum, A.J.; Jefferies, J.R.; Barrett, J.; Brophy, P.M.; Ferreira, H.B.; Zaha, A. Proteomic analysis of the larval stage of the parasite Echinococcus granulosus: Causative agent of cystic hydatid disease. *Proteomics* **2003**, *3*, 1633–1636. [CrossRef] [PubMed]
89. Kouguchi, H.; Matsumoto, J.; Katoh, Y.; Suzuki, T.; Oku, Y.; Yagi, K. Echinococcus multilocularis: Two-dimensional Western blotting method for the identification and expression analysis of immunogenic proteins in infected dogs. *Exp. Parasitol.* **2010**, *124*, 238–243. [CrossRef] [PubMed]
90. Wang, Y.; Cheng, Z.; Lu, X.; Tang, C. Echinococcus multilocularis: Proteomic analysis of the protoscoleces by two-dimensional electrophoresis and mass spectrometry. *Exp. Parasitol.* **2009**, *123*, 162–167. [CrossRef]
91. Hidalgo, C.; García, M.P.; Stoore, C.; Ramírez, J.P.; Monteiro, K.M.; Hellman, U.; Zaha, A.; Ferreira, H.B.; Galanti, N.; Landerer, E.; et al. Proteomics analysis of *Echinococcus granulosus* protoscolex stage. *Vet. Parasitol.* **2016**, *218*, 43–45. [CrossRef] [PubMed]
92. Manterola, C.; García, N.; Rojas, C. Aspectos Generales del Perfil Proteómico del *Echinococcus granulosus*. *Int. J. Morphol.* **2019**, *37*, 773–779. [CrossRef]
93. Miles, S.; Portela, M.; Cyrklaff, M.; Ancarola, M.E.; Frischknecht, F.; Durán, R.; Dematteis, S.; Mourglia-Ettlin, G. Combining proteomics and bioinformatics to explore novel tegumental antigens as vaccine candidates against *Echinococcus granulosus* infection. *J. Cell. Biochem.* **2019**, *120*, 15320–15336. [CrossRef]
94. Monteiro, K.M.; De Carvalho, M.O.; Zaha, A.; Ferreira, H.B. Proteomic analysis of the *Echinococcus granulosus* metacestode during infection of its intermediate host. *Proteomics* **2010**, *10*, 1985–1999. [CrossRef]
95. Longuespée, R.; Casadonte, R.; Kriegsmann, M.; Wandernoth, P.; Lisenko, K.; Mazzucchelli, G.; Becker, M.; Kriegsmann, J. Proteomic investigation of human cystic echinococcosis in the liver. *Mol. Biochem. Parasitol.* **2017**, *211*, 9–14. [CrossRef]
96. Ahn, C.S.; Kim, J.G.; Han, X.; Kang, I.; Kong, Y. Comparison of *Echinococcus multilocularis* and *Echinococcus granulosus* hydatid fluid proteome provides molecular strategies for specialized host-parasite interactions. *Oncotarget* **2017**, *8*, 97009–97024. [CrossRef]
97. Cui, S.J.; Xu, L.L.; Zhang, T.; Xu, M.; Yao, J.; Fang, C.Y.; Feng, Z.; Yang, P.Y.; Hu, W.; Liu, F. Proteomic characterization of larval and adult developmental stages in Echinococcus granulosus reveals novel insight into host-parasite interactions. *J. Proteom.* **2013**, *84*, 158–175. [CrossRef]
98. Zhou, X.; Wang, W.; Cui, F.; Shi, C.; Ma, Y.; Yu, Y.; Zhao, W.; Zhao, J. Extracellular vesicles derived from *Echinococcus granulosus* hydatid cyst fluid from patients: Isolation, characterization and evaluation of immunomodulatory functions on T cells. *Int. J. Parasitol.* **2019**, *49*, 1029–1037. [CrossRef] [PubMed]
99. Rostami, A.; Ma, G.; Wang, T.; Koehler, A.V.; Hofmann, A.; Chang, B.C.H.; Macpherson, C.N.; Gasser, R.B. Human toxocariasis—A look at a neglected disease through an epidemiological 'prism'. *Infect. Genet. Evol.* **2019**, *74*, 104002. [CrossRef] [PubMed]
100. Schnieder, T.; Laabs, E.M.; Welz, C. Larval development of *Toxocara canis* in dogs. *Vet. Parasitol.* **2011**, *175*, 193–206. [CrossRef] [PubMed]
101. Zhu, X.Q.; Korhonen, P.K.; Cai, H.; Young, N.D.; Nejsum, P.; Von Samson-Himmelstjerna, G.; Boag, P.R.; Tan, P.; Li, Q.; Min, J.; et al. Genetic blueprint of the zoonotic pathogen *Toxocara canis*. *Nat. Commun.* **2015**, *6*. [CrossRef] [PubMed]
102. Da Silva, M.B.; Oviedo, Y.; Cooper, P.J.; Pacheco, L.G.C.; Pinheiro, C.S.; Ferreira, F.; Briza, P.; Alcantara-Neves, N.M. The somatic proteins of *Toxocara canis* larvae and excretory-secretory products revealed by proteomics. *Vet. Parasitol.* **2018**, *259*, 25–34. [CrossRef]
103. Sperotto, R.L.; Kremer, F.S.; Aires Berne, M.E.; Costa de Avila, L.F.; da Silva Pinto, L.; Monteiro, K.M.; Caumo, K.S.; Ferreira, H.B.; Berne, N.; Borsuk, S. Proteomic analysis of *Toxocara canis* excretory and secretory (TES) proteins. *Mol. Biochem. Parasitol.* **2017**, *211*, 39–47. [CrossRef]

104. Almagro Armenteros, J.J.; Tsirigos, K.D.; Sønderby, C.K.; Petersen, T.N.; Winther, O.; Brunak, S.; von Heijne, G.; Nielsen, H. SignalP 5.0 improves signal peptide predictions using deep neural networks. *Nat. Biotechnol.* **2019**, *37*, 420–423. [CrossRef]
105. Dold, C.; Holland, C.V. Ascaris and ascariasis. *Microbes Infect.* **2011**, *13*, 632–637. [CrossRef]
106. De Silva, N.R.; Brooker, S.; Hotez, P.J.; Montresor, A.; Engels, D.; Savioli, L. Soil-transmitted helminth infections: Updating the global picture. *Trends Parasitol.* **2003**, *19*, 547–551. [CrossRef]
107. Mrozińska-Gogol, J. Ascaris lumbricoides. In *Medical Parasitology [Parazytologia medyczna]*; Wydawnictwo Lekarskie PZWL: Warsaw, Poland, 2016; pp. 174–180. ISBN 978-83-200-5138-4.
108. Xu, M.J.; Fu, J.H.; Zhou, D.H.; Elsheikha, H.M.; Hu, M.; Lin, R.Q.; Peng, L.F.; Song, H.Q.; Zhu, X.Q. *Ascaris lumbricoides* and *Ascaris suum*: Comparative proteomic studies using 2-DE coupled with mass spectrometry. *Int. J. Mass Spectrom.* **2013**, *339*, 1–6. [CrossRef]
109. Abebe, W.; Tsuji, N.; Kasuga-Aoki, H.; Miyoshi, T.; Isobe, T.; Arakawa, T.; Matsumoto, Y.; Yoshihara, S. Species-specific proteins identified in Ascaris lumbricoides and Ascaris suum using two-dimensional electrophoresis. *Parasitol. Res.* **2002**, *88*, 868–871. [CrossRef] [PubMed]
110. Nagorny, S.A.; Aleshukina, A.V.; Aleshukina, I.S.; Ermakova, L.A.; Pshenichnaya, N.Y. The application of proteomic methods (MALDI-toff MS) for studying protein profiles of some nematodes (dirofilaria and ascaris) for differentiating species. *Int. J. Infect. Dis.* **2019**, *82*, 61–65. [CrossRef]
111. González-Miguel, J.; Morchón, R.; Gussoni, S.; Bossetti, E.; Hormaeche, M.; Kramer, L.H.; Simón, F. Immunoproteomic approach for identification of Ascaris suum proteins recognized by pigs with porcine ascariasis. *Vet. Parasitol.* **2014**, *203*, 343–348. [CrossRef] [PubMed]
112. Chehayeb, J.F.; Robertson, A.P.; Martin, R.J.; Geary, T.G. Proteomic Analysis of Adult Ascaris suum Fluid Compartments and Secretory Products. *PLoS Negl. Trop. Dis.* **2014**, *8*. [CrossRef] [PubMed]
113. Hansen, E.P.; Fromm, B.; Andersen, S.D.; Marcilla, A.; Andersen, K.L.; Borup, A.; Williams, A.R.; Jex, A.R.; Gasser, R.B.; Young, N.D.; et al. Exploration of extracellular vesicles from Ascaris suum provides evidence of parasite–host cross talk. *J. Extracell. Vesicles* **2019**, *8*. [CrossRef]
114. Robinson, M.W.; Dalton, J.P. Zoonotic helminth infections with particular emphasis on fasciolosis and other trematodiases. *Philos. Trans. R. Soc. B Biol. Sci.* **2009**, *364*, 2763–2776. [CrossRef]
115. Mas-Coma, S.; Bargues, M.D.; Valero, M.A. Fascioliasis and other plant-borne trematode zoonoses. *Int. J. Parasitol.* **2005**, *35*, 1255–1278. [CrossRef]
116. Mas-Coma, S.; Bargues, M.D.; Valero, M.A. Plant-Borne Trematode Zoonoses: Fascioliasis and Fasciolopsiasis. In *Food-Borne Parasitic Zoonoses: Fish and Plant-Borne Parasites*; Murrell, D.K., Fried, B., Eds.; Springer: New York, NY, USA, 2007; pp. 293–334.
117. Mas-Coma, S. Human Fascoliasis: Epidemiological patterns in human endemic areas of South America, Africa and Asia. *Southeast Asian J. Trop. Med. Public Health* **2004**, *35*, 1–11.
118. Keiser, J.; Utzinger, J. Emerging Foodborne Trematodiasis. *Emerg. Infect. Dis.* **2005**, *11*, 1507–1514. [CrossRef]
119. Cwiklinski, K.; O'Neill, S.M.; Donnelly, S.; Dalton, J.P. A prospective view of animal and human Fasciolosis. *Parasite Immunol.* **2016**, *38*, 558–568. [CrossRef]
120. Irving, D.O.; Howell, M.J. Characterization of excretory-secretory antigens of *Fasciola hepatica*. *Parasitology* **1982**, *85*, 179–188. [CrossRef] [PubMed]
121. Dalton, J.P.; Tom, T.D.; Strand, M. *Fasciola hepatica*: Comparison of immature and mature immunoreactive glycoproteins. *Parasite Immunol.* **1985**, *7*, 643–657. [CrossRef] [PubMed]
122. Lee, C.G.; Zimmerman, G.L.; Bishop, J.K. Host influence on the banding profiles of whole-body protein and excretory-secretory product of *Fasciola hepatica* (trematoda) by isoelectric focusing. *Vet. Parasitol.* **1992**, *41*, 57–68. [CrossRef]
123. Lee, C.G.; Zimmerman, G.L.; Mulrooney, D.M. Isoelectric focusing of soluble proteins from *Fasciola hepatica* L, 1758 and Fascioloides magna B, 1875. *Am. J. Vet. Res.* **1992**, *53*, 246–250. [PubMed]
124. Jefferies, J.R.; Brophy, P.M.; Barrett, J. Investigation of *Fasciola hepatica* sample preparation for two-dimensional electrophoresis. *Electrophoresis* **2000**, *21*, 3724–3729. [CrossRef]
125. Jefferies, J.R.; Campbell, A.M.; van Rossum, A.J.; Barrett, J.; Brophy, P.M. Proteomic analysis of *Fasciola hepatica* excretory-secretory products. *Proteomics* **2001**, *1*, 1128–1132. [CrossRef]
126. Chemale, G.; Morphew, R.; Moxon, J.V.; Morassuti, A.L.; LaCourse, E.J.; Barrett, J.; Johnston, D.A.; Brophy, P.M. Proteomic analysis of glutathione transferases from the liver fluke parasite, *Fasciola hepatica*. *Proteomics* **2006**, *6*, 6263–6273. [CrossRef]

127. Robinson, M.W.; Tort, J.F.; Lowther, J.; Donnelly, S.M.; Wong, E.; Xu, W.; Stack, C.M.; Padula, M.; Herbert, B.; Dalton, J.P. Proteomics and Phylogenetic Analysis of the Cathepsin L Protease Family of the Helminth Pathogen *Fasciola hepatica*. *Mol. Cell. Proteom.* **2008**, *7*, 1111–1123. [CrossRef]
128. Marcilla, A.; De la Rubia, J.E.; Sotillo, J.; Bernal, D.; Carmona, C.; Villavicencio, Z.; Acosta, D.; Tort, J.; Bornay, F.J.; Esteban, J.G.; et al. Leucine Aminopeptidase Is an Immunodominant Antigen of *Fasciola hepatica* Excretory and Secretory Products in Human Infections. *Clin. Vaccine Immunol.* **2008**, *15*, 95–100. [CrossRef]
129. Morphew, R.M.; Wright, H.A.; LaCourse, E.J.; Porter, J.; Barrett, J.; Woods, D.J.; Brophy, P.M. Towards Delineating Functions within the Fasciola Secreted Cathepsin L Protease Family by Integrating In Vivo Based Sub-Proteomics and Phylogenetics. *PLoS Negl. Trop. Dis.* **2011**, *5*, e937. [CrossRef]
130. Morphew, R.M.; Eccleston, N.; Wilkinson, T.J.; McGarry, J.; Perally, S.; Prescott, M.; Ward, D.; Williams, D.; Paterson, S.; Raman, M.; et al. Proteomics and in Silico Approaches To Extend Understanding of the Glutathione Transferase Superfamily of the Tropical Liver Fluke Fasciola gigantica. *J. Proteome Res.* **2012**, *11*, 5876–5889. [CrossRef]
131. Morphew, R.M.; Hamilton, C.M.; Wright, H.A.; Dowling, D.J.; O'Neill, S.M.; Brophy, P.M. Identification of the major proteins of an immune modulating fraction from adult *Fasciola hepatica* released by Nonidet P40. *Vet. Parasitol.* **2013**, *191*, 379–385. [CrossRef] [PubMed]
132. Morphew, R.M.; Wilkinson, T.J.; Mackintosh, N.; Jahndel, V.; Paterson, S.; McVeigh, P.; Abbas Abidi, S.M.; Saifullah, K.; Raman, M.; Ravikumar, G.; et al. Exploring and Expanding the Fatty-Acid-Binding Protein Superfamily in Fasciola Species. *J. Proteome Res.* **2016**, *15*, 3308–3321. [CrossRef] [PubMed]
133. Cwiklinski, K.; de la Torre-Escudero, E.; Trelis, M.; Bernal, D.; Dufresne, P.J.; Brennan, G.P.; O'Neill, S.; Tort, J.; Paterson, S.; Marcilla, A.; et al. The Extracellular Vesicles of the Helminth Pathogen, *Fasciola hepatica*: Biogenesis Pathways and Cargo Molecules Involved in Parasite Pathogenesis. *Mol. Cell. Proteom.* **2015**, *14*, 3258–3273. [CrossRef] [PubMed]
134. Di Maggio, L.S.; Tirloni, L.; Pinto, A.F.M.; Diedrich, J.K.; Yates III, J.R.; Benavides, U.; Carmona, C.; da Silva Vaz, I.; Berasain, P. Across intra-mammalian stages of the liver f luke *Fasciola hepatica*: A proteomic study. *Sci. Rep.* **2016**, *6*, 32796. [CrossRef]
135. Chemale, G.; Perally, S.; LaCourse, E.J.; Prescott, M.C.; Jones, L.M.; Ward, D.; Meaney, M.; Hoey, E.; Brennan, G.P.; Fairweather, I.; et al. Comparative Proteomic Analysis of Triclabendazole Response in the Liver Fluke Fasciola hepatica. *J. Proteome Res.* **2010**, *9*, 4940–4951. [CrossRef]
136. Morphew, R.M.; MacKintosh, N.; Hart, E.H.; Prescott, M.; LaCourse, E.J.; Brophy, P.M. In vitro biomarker discovery in the parasitic flatworm Fasciola hepatica for monitoring chemotherapeutic treatment. *EuPA Open Proteom.* **2014**, *3*, 85–99. [CrossRef]
137. Moxon, J.V.; LaCourse, E.J.; Wright, H.A.; Perally, S.; Prescott, M.C.; Gillard, J.L.; Barrett, J.; Hamilton, J.V.; Brophy, P.M. Proteomic analysis of embryonic Fasciola hepatica: Characterization and antigenic potential of a developmentally regulated heat shock protein. *Vet. Parasitol.* **2010**, *169*, 62–75. [CrossRef]
138. Wilson, R.A.; Wright, J.M.; de Castro-Borges, W.; Parker-Manuel, S.J.; Dowle, A.A.; Ashton, P.D.; Young, N.D.; Gasser, R.B.; Spithill, T.W. Exploring the Fasciola hepatica tegument proteome. *Int. J. Parasitol.* **2011**, *41*, 1347–1359. [CrossRef]
139. Haçarız, O.; Sayers, G.; Baykal, A.T. A Proteomic Approach To Investigate the Distribution and Abundance of Surface and Internal Fasciola hepatica Proteins during the Chronic Stage of Natural Liver Fluke Infection in Cattle. *J. Proteome Res.* **2012**, *11*, 3592–3604. [CrossRef]
140. Ley, V.; Andrews, N.W.; Robbins, E.S.; Nussenzweig, V. Amastigotes of Trypanosoma cruzi sustain an infective cycle in mammalian cells. *J. Exp. Med.* **1988**, *168*, 649–659. [CrossRef] [PubMed]
141. Monteiro, F.A.; Weirauch, C.; Felix, M.; Lazoski, C.; Abad-Franch, F. Evolution, Systematics, and Biogeography of the Triatominae, Vectors of Chagas Disease. In *Advances in Parasitology*; Elsevier: Amsterdam, The Netherlands, 2018; pp. 265–344.
142. Berry, A.S.F.; Salazar-Sánchez, R.; Castillo-Neyra, R.; Borrini-Mayorí, K.; Chipana-Ramos, C.; Vargas-Maquera, M.; Ancca-Juarez, J.; Náquira-Velarde, C.; Levy, M.Z.; Brisson, D. Sexual reproduction in a natural Trypanosoma cruzi population. *PLoS Negl. Trop. Dis.* **2019**, *13*, e0007392. [CrossRef] [PubMed]
143. Pérez-Molina, J.A.; Molina, I. Chagas disease. *Lancet* **2018**, *391*, 82–94. [CrossRef]
144. Bern, C. Chagas' Disease. *N. Engl. J. Med.* **2015**, *373*, 456–466. [CrossRef] [PubMed]
145. Atwood, J.A.; Weatherly, D.B.; Minning, T.A.; Bundy, B.; Cavola, C.; Opperdoes, F.R.; Orlando, R.; Tarleton, R.L. The Trypanosoma cruzi Proteome. *Science* **2005**, *309*, 473–476. [CrossRef]

146. Brunoro, G.V.F.; Caminha, M.A.; Ferreira, A.T. da S.; da Veiga Leprevost, F.; Carvalho, P.C.; Perales, J.; Valente, R.H.; Menna-Barreto, R.F.S. Reevaluating the Trypanosoma cruzi proteomic map: The shotgun description of bloodstream trypomastigotes. *J. Proteom.* **2015**, *115*, 58–65. [CrossRef]
147. Nakayasu, E.S.; Sobreira, T.J.P.; Torres, R.; Ganiko, L.; Oliveira, P.S.L.; Marques, A.F.; Almeida, I.C. Improved Proteomic Approach for the Discovery of Potential Vaccine Targets in *Trypanosoma cruzi*. *J. Proteome Res.* **2012**, *11*, 237–246. [CrossRef]
148. Parodi-Talice, A.; Monteiro-Goes, V.; Arrambide, N.; Avila, A.R.; Duran, R.; Correa, A.; Dallagiovanna, B.; Cayota, A.; Krieger, M.; Goldenberg, S.; et al. Proteomic analysis of metacyclic trypomastigotes undergoing *Trypanosoma cruzi* metacyclogenesis. *J. Mass Spectrom.* **2007**, *42*, 1422–1432. [CrossRef]
149. Amorim, J.C.; Batista, M.; da Cunha, E.S.; Lucena, A.C.R.; de Paula Lima, C.V.; Sousa, K.; Krieger, M.A.; Marchini, F.K. Quantitative proteome and phosphoproteome analyses highlight the adherent population during *Trypanosoma cruzi* metacyclogenesis. *Sci. Rep.* **2017**, *7*, 9899. [CrossRef]
150. Lucena, A.C.R.; Amorim, J.C.; de Paula Lima, C.V.; Batista, M.; Krieger, M.A.; de Godoy, L.M.F.; Marchini, F.K. Quantitative phosphoproteome and proteome analyses emphasize the influence of phosphorylation events during the nutritional stress of *Trypanosoma cruzi*: The initial moments of in vitro metacyclogenesis. *Cell Stress Chaperones* **2019**, *24*, 927–936. [CrossRef]
151. Avila, C.; Mule, S.; Rosa-Fernandes, L.; Viner, R.; Barisón, M.; Costa-Martins, A.; Oliveira, G.; Teixeira, M.; Marinho, C.; Silber, A.; et al. Proteome-Wide Analysis of *Trypanosoma cruzi* Exponential and Stationary Growth Phases Reveals a Subcellular Compartment-Specific Regulation. *Genes* **2018**, *9*, 413. [CrossRef]
152. Kessler, R.L.; Contreras, V.T.; Marliére, N.P.; Aparecida Guarneri, A.; Villamizar Silva, L.H.; Mazzarotto, G.A.C.A.; Batista, M.; Soccol, V.T.; Krieger, M.A.; Probst, C.M. Recently differentiated epimastigotes from *Trypanosoma cruzi* are infective to the mammalian host. *Mol. Microbiol.* **2017**, *104*, 712–736. [CrossRef] [PubMed]
153. Queiroz, R.M.L.; Charneau, S.; Mandacaru, S.C.; Schwämmle, V.; Lima, B.D.; Roepstorff, P.; Ricart, C.A.O. Quantitative Proteomic and Phosphoproteomic Analysis of *Trypanosoma cruzi* Amastigogenesis. *Mol. Cell. Proteom.* **2014**, *13*, 3457–3472. [CrossRef] [PubMed]
154. Martins, N.O.; de Souza, R.T.; Cordero, E.M.; Maldonado, D.C.; Cortez, C.; Marini, M.M.; Ferreira, E.R.; Bayer-Santos, E.; de Almeida, I.C.; Yoshida, N.; et al. Molecular Characterization of a Novel Family of *Trypanosoma cruzi* Surface Membrane Proteins (TcSMP) Involved in Mammalian Host Cell Invasion. *PLoS Negl. Trop. Dis.* **2015**, *9*, e0004216. [CrossRef] [PubMed]
155. Hayes, K.S.; Bancroft, A.J.; Goldrick, M.; Portsmouth, C.; Roberts, I.S.; Grencis, R.K. Exploitation of the Intestinal Microflora by the Parasitic Nematode Trichuris muris. *Science* **2010**, *328*, 1391–1394. [CrossRef] [PubMed]
156. Centers for Disease Control and Prevention Parasites—Trichuriasis (Also Known as Whipworm Infection). Available online: https://www.cdc.gov/parasites/whipworm/ (accessed on 22 July 2020).
157. Lillywhite, J.E.; Cooper, E.S.; Needham, C.S.; Venugopal, S.; Bundy, D.A.P.; Bianco, A.E. Identification and characterization of excreted/secreted products of *Trichuris trichiura*. *Parasite Immunol.* **1995**, *17*, 47–54. [CrossRef] [PubMed]
158. Cruz, K.; Marcilla, P.; Kelly, P.; Vandenplas, M.; Osuna, A.; Trelis, M. Proteomic Analysis of Trichuris Trichiura Egg Extract Reveals Potential Immunomodulators and Diagnostic Targets. *Res. Sq.* **2020**. [CrossRef]
159. Hurst, R.J.M.; Else, K.J. *Trichuris muris* research revisited: A journey through time. *Parasitology* **2013**, *140*, 1325–1339. [CrossRef]
160. Eichenberger, R.M.; Talukder, M.H.; Field, M.A.; Wangchuk, P.; Giacomin, P.; Loukas, A.; Sotillo, J. Characterization of *Trichuris muris* secreted proteins and extracellular vesicles provides new insights into host–parasite communication. *J. Extracell. Vesicles* **2018**, *7*, 1428004. [CrossRef]
161. Tritten, L.; Tam, M.; Vargas, M.; Jardim, A.; Stevenson, M.M.; Keiser, J.; Geary, T.G. Excretory/secretory products from the gastrointestinal nematode *Trichuris muris*. *Exp. Parasitol.* **2017**, *178*, 30–36. [CrossRef]
162. Shears, R.K.; Bancroft, A.J.; Sharpe, C.; Grencis, R.K.; Thornton, D.J. Vaccination Against Whipworm: Identification of Potential Immunogenic Proteins in *Trichuris muris* Excretory/Secretory Material. *Sci. Rep.* **2018**, *8*, 4508. [CrossRef] [PubMed]
163. Dubey, J.P. *Toxoplasmosis of Animals and Humans*; CRC Press: Boca Raton, FL, USA, 2016; Volume 3, ISBN 9780429092954.

164. Kijlstra, A.; Jongert, E. Toxoplasma-safe meat: Close to reality? *Trends Parasitol.* **2009**, *25*, 18–22. [CrossRef] [PubMed]
165. Cohen, A.M.; Rumpel, K.; Coombs, G.H.; Wastling, J.M. Characterisation of global protein expression by two-dimensional electrophoresis and mass spectrometry: Proteomics of Toxoplasma gondii. *Int. J. Parasitol.* **2002**, *32*, 39–51. [CrossRef]
166. Nischik, N.; Schade, B.; Dytnerska, K.; Długońska, H.; Reichmann, G.; Fischer, H.G. Attenuation of mouse-virulent Toxoplasma gondii parasites is associated with a decrease in interleukin-12-inducing tachyzoite activity and reduced expression of actin, catalase and excretory proteins. *Microbes Infect.* **2001**, *3*, 689–699. [CrossRef]
167. Wastling, J.M.; Xia, D. Proteomics of *Toxoplasma gondii*. In *Toxoplasma Gondii: The Model Apicomplexan—Perspectives and Methods: Second Edition*; Elsevier: Amsterdam, The Netherlands, 2013; pp. 731–754. ISBN 9780123964816.
168. Xia, D.; Sanderson, S.J.; Jones, A.R.; Prieto, J.H.; Yates, J.R.; Bromley, E.; Tomley, F.M.; Lal, K.; Sinden, R.E.; Brunk, B.P.; et al. The proteome of *Toxoplasma gondii*: Integration with the genome provides novel insights into gene expression and annotation. *Genome Biol.* **2008**, *9*, R116. [CrossRef]
169. Dybas, J.M.; Madrid-Aliste, C.J.; Che, F.Y.; Nieves, E.; Rykunov, D.; Angeletti, R.H.; Weiss, L.M.; Kim, K.; Fiser, A. Computational Analysis and Experimental Validation of Gene Predictions in Toxoplasma gondii. *PLoS ONE* **2008**, *3*, e3899. [CrossRef]
170. Zhou, D.H.; Zhao, F.R.; Nisbet, A.J.; Xu, M.J.; Song, H.Q.; Lin, R.Q.; Huang, S.Y.; Zhu, X.Q. Comparative proteomic analysis of different *Toxoplasma gondii* genotypes by two-dimensional fluorescence difference gel electrophoresis combined with mass spectrometry. *Electrophoresis* **2014**, *35*, 533–545. [CrossRef]
171. Ma, G.Y.; Zhang, J.Z.; Yin, G.R.; Zhang, J.H.; Meng, X.L.; Zhao, F. *Toxoplasma gondii*: Proteomic analysis of antigenicity of soluble tachyzoite antigen. *Exp. Parasitol.* **2009**, *122*, 41–46. [CrossRef]
172. Krishna, R.; Xia, D.; Sanderson, S.; Shanmugasundram, A.; Vermont, S.; Bernal, A.; Daniel-Naguib, G.; Ghali, F.; Brunk, B.P.; Roos, D.S.; et al. A large-scale proteogenomics study of apicomplexan pathogens-*Toxoplasma gondii* and *Neospora caninum*. *Proteomics* **2015**, *15*, 2618–2628. [CrossRef]
173. Fritz, H.M.; Bowyer, P.W.; Bogyo, M.; Conrad, P.A.; Boothroyd, J.C. Proteomic Analysis of Fractionated Toxoplasma Oocysts Reveals Clues to Their Environmental Resistance. *PLoS ONE* **2012**, *7*, e29955. [CrossRef]
174. Zhou, C.X.; Zhu, X.Q.; Elsheikha, H.M.; He, S.; Li, Q.; Zhou, D.H.; Suo, X. Global iTRAQ-based proteomic profiling of *Toxoplasma gondii* oocysts during sporulation. *J. Proteom.* **2016**, *148*, 12–19. [CrossRef] [PubMed]
175. Zhou, X.W.; Kafsack, B.F.C.; Cole, R.N.; Beckett, P.; Shen, R.F.; Carruthers, V.B. The Opportunistic Pathogen *Toxoplasma gondii* Deploys a Diverse Legion of Invasion and Survival Proteins. *J. Biol. Chem.* **2005**, *280*, 34233–34244. [CrossRef] [PubMed]
176. Lee, W.K.; Ahn, H.J.; Baek, J.H.; Lee, C.H.; Yu, Y.G.; Nam, H.W. Comprehensive proteome analysis of the excretory/secretory proteins of toxoplasma gondii. *Bull. Korean Chem. Soc.* **2014**, *35*, 3071–3076. [CrossRef]
177. Barylyuk, K.; Koreny, L.; Ke, H.; Butterworth, S.; Crook, O.M.; Lassadi, I.; Gupta, V.; Tromer, E.; Mourier, T.; Stevens, T.J.; et al. A subcellular atlas of Toxoplasma reveals the functional context of the proteome. *bioRxiv* **2020**. [CrossRef]
178. Christoforou, A.; Mulvey, C.M.; Breckels, L.M.; Geladaki, A.; Hurrell, T.; Hayward, P.C.; Naake, T.; Gatto, L.; Viner, R.; Arias, A.M.; et al. A draft map of the mouse pluripotent stem cell spatial proteome. *Nat. Commun.* **2016**, *7*, 9992. [CrossRef]
179. Mulvey, C.M.; Breckels, L.M.; Geladaki, A.; Britovšek, N.K.; Nightingale, D.J.H.; Christoforou, A.; Elzek, M.; Deery, M.J.; Gatto, L.; Lilley, K.S. Using hyperLOPIT to perform high-resolution mapping of the spatial proteome. *Nat. Protoc.* **2017**, *12*, 1110–1135. [CrossRef]
180. Pozio, E.; Darwin Murrell, K. Systematics and Epidemiology of Trichinella. *Adv. Parasitol.* **2006**, *63*, 367–439.
181. Gottstein, B.; Pozio, E.; Nöckler, K. Epidemiology, Diagnosis, Treatment, and Control of Trichinellosis. *Clin. Microbiol. Rev.* **2009**, *22*, 127–145. [CrossRef]
182. Liu, J.Y.; Zhang, N.Z.; Li, W.H.; Li, L.; Yan, H.B.; Qu, Z.G.; Li, T.T.; Cui, J.M.; Yang, Y.; Jia, W.Z.; et al. Proteomic analysis of differentially expressed proteins in the three developmental stages of *Trichinella spiralis*. *Vet. Parasitol.* **2016**, *231*, 32–38. [CrossRef]
183. Ren, H.N.; Liu, R.D.; Song, Y.Y.; Zhuo, T.X.; Guo, K.X.; Zhang, Y.; Jiang, P.; Wang, Z.Q.; Cui, J. Label-free quantitative proteomic analysis of molting-related proteins of *Trichinella spiralis* intestinal infective larvae. *Vet. Res.* **2019**, *50*, 70. [CrossRef]

184. Grzelak, S.; Moskwa, B.; Bień, J. *Trichinella britovi* muscle larvae and adult worms: Stage-specific and common antigens detected by two-dimensional gel electrophoresis-based immunoblotting 06 Biological Sciences 0601 Biochemistry and Cell Biology. *Parasites Vectors* **2018**, *11*, 1–17. [CrossRef] [PubMed]
185. Dea-Ayuela, M.A.; Bolás-Fernández, F. Two-dimensional electrophoresis and mass spectrometry for the identification of species-specific Trichinella antigens. *Vet. Parasitol.* **2005**, *132*, 43–49. [CrossRef] [PubMed]
186. Yang, J.; Pan, W.; Sun, X.; Zhao, X.; Yuan, G.; Sun, Q.; Huang, J.; Zhu, X. Immunoproteomic profile of *Trichinella spiralis* adult worm proteins recognized by early infection sera. *Parasites Vectors* **2015**, *8*, 20. [CrossRef] [PubMed]
187. Somboonpatarakun, C.; Rodpai, R.; Intapan, P.M.; Sanpool, O.; Sadaow, L.; Wongkham, C.; Insawang, T.; Boonmars, T.; Maleewong, W. Immuno-proteomic analysis of *Trichinella spiralis*, *T. pseudospiralis*, and *T. papuae* extracts recognized by human *T. spiralis*-infected sera. *Parasitol. Res.* **2018**, *117*, 201–212. [CrossRef]
188. Liu, R.D.; Cui, J.; Wang, L.; Al, E. Identification of surface proteins of *Trichinella spiralis* muscle larvae using immunoproteomics. *Trop. Biomed.* **2014**, *31*, 579–591.
189. Liu, R.D.; Cui, J.; Liu, X.L.; Jiang, P.; Sun, G.G.; Zhang, X.; Long, S.R.; Wang, L.; Wang, Z.Q. Comparative proteomic analysis of surface proteins of *Trichinella spiralis* muscle larvae and intestinal infective larvae. *Acta Trop.* **2015**, *150*, 79–86. [CrossRef] [PubMed]
190. Wang, Y.; Bai, X.; Zhu, H.; Wang, X.; Shi, H.; Tang, B.; Boireau, P.; Cai, X.; Luo, X.; Liu, M.; et al. Immunoproteomic analysis of the excretory-secretory products of *Trichinella pseudospiralis* adult worms and newborn larvae. *Parasites Vectors* **2017**, *10*, 579. [CrossRef] [PubMed]
191. Grzelak, S.; Stachyra, A.; Bień-Kalinowska, J. The first analysis of *Trichinella spiralis* and *Trichinella britovi* adult worm excretory-secretory proteins by two-dimensional electrophoresis coupled with LC-MS/MS. *Vet. Parasitol.* **2020**, 109096. [CrossRef]
192. Wang, Y.; Bai, X.; Tang, B.; Zhang, Y.; Zhang, L.; Cai, X.; Lin, J.; Jia, W.; Boireau, P.; Liu, M.; et al. Comparative analysis of excretory–secretory products of muscle larvae of three isolates of Trichinella pseudospiralis by the iTRAQ method. *Vet. Parasitol.* **2020**, 109119. [CrossRef]
193. Djurković-Djaković, O.; Bobić, B.; Nikolić, A.; Klun, I.; Dupouy-Camet, J. Pork as a source of human parasitic infection. *Clin. Microbiol. Infect.* **2013**, *19*, 586–594. [CrossRef]
194. Dorny, P.; Vallée, I.; Alban, L.; Boes, J.; Boireau, P.; Boué, F.; Claes, M.; Cook, A.J.C.; Enemark, H.; van der Giessen, J.; et al. Development of harmonised schemes for the monitoring and reporting of Cysticercus in animals and foodstuffs in the European Union. *EFSA Support. Publ.* **2010**, *7*. [CrossRef]
195. García, H.H.; Gonzalez, A.E.; Evans, C.A.; Gilman, R.H. Taenia solium cysticercosis. *Lancet* **2003**, *362*, 547–556. [CrossRef]
196. Fang, W.; Xiao, L.L.; Bao, H.E.; Mu, R. Total protein analysis by two-dimensional electrophoresis in cysticerci of Taenia solium and *Taenia asiatica*. *Zhongguo Ji Sheng Chong Xue Yu Ji Sheng Chong Bing Za Zhi* **2011**, *29*, 188–190. [PubMed]
197. Santivañez, S.J.; Hernández-González, A.; Chile, N.; Oleaga, A.; Arana, Y.; Palma, S.; Verastegui, M.; Gonzalez, A.E.; Gilman, R.; Garcia, H.H.; et al. Proteomic study of activated *Taenia solium* oncospheres. *Mol. Biochem. Parasitol.* **2010**, *171*, 32–39. [CrossRef] [PubMed]
198. Diaz-Masmela, Y.; Fragoso, G.; Ambrosio, J.R.; Mendoza-Hernández, G.; Rosas, G.; Estrada, K.; Carrero, J.C.; Sciutto, E.; Laclette, J.P.; Bobes, R.J. Immunodiagnosis of porcine cysticercosis: Identification of candidate antigens through immunoproteomics. *Vet. J.* **2013**, *198*, 656–660. [CrossRef]
199. Esquivel-Velázquez, M.; Larralde, C.; Morales, J.; Ostoa-Saloma, P. Protein and antigen diversity in the vesicular fluid of taenia solium cysticerci dissected from naturally infected pigs. *Int. J. Biol. Sci.* **2011**, *7*, 1287–1297. [CrossRef]
200. Navarrete-Perea, J.; Moguel, B.; Mendoza-Hernández, G.; Fragoso, G.; Sciutto, E.; Bobes, R.J.; Laclette, J.P. Identification and quantification of host proteins in the vesicular fluid of porcine *Taenia solium* cysticerci. *Exp. Parasitol.* **2014**, *143*, 11–17. [CrossRef]
201. Bae, Y.A.; Yeom, J.S.; Wang, H.; Kim, S.H.; Ahn, C.S.; Kim, J.T.; Yang, H.J.; Kong, Y. *Taenia solium* metacestode fasciclin-like protein is reactive with sera of chronic neurocysticercosis. *Trop. Med. Int. Health* **2014**, *19*, 719–725. [CrossRef]
202. Navarrete-Perea, J.; Moguel, B.; Bobes, R.J.; Villalobos, N.; Carrero, J.C.; Sciutto, E.; Soberón, X.; Laclette, J.P. Protein profiles of *Taenia solium* cysts obtained from skeletal muscles and the central nervous system of pigs: Search for tissue-specific proteins. *Exp. Parasitol.* **2017**, *172*, 23–29. [CrossRef]

203. Navarrete-Perea, J.; Isasa, M.; Paulo, J.A.; Corral-Corral, R.; Flores-Bautista, J.; Hernández-Téllez, B.; Bobes, R.J.; Fragoso, G.; Sciutto, E.; Soberón, X.; et al. Quantitative multiplexed proteomics of *Taenia solium* cysts obtained from the skeletal muscle and central nervous system of pigs. *PLoS Negl. Trop. Dis.* **2017**, *11*, e0005962. [CrossRef]
204. da Costa, G.C.V.; Peralta, R.H.S.; Kalume, D.E.; Alves, A.L.G.M.; Peralta, J.M. A gel-free proteomic analysis of *Taenia solium* and *Taenia crassiceps* cysticerci vesicular extracts. *Parasitol. Res.* **2018**, *117*, 3781–3790. [CrossRef] [PubMed]
205. Victor, B.; Kanobana, K.; Gabriël, S.; Polman, K.; Deckers, N.; Dorny, P.; Deelder, A.M.; Palmblad, M. Proteomic analysis of *Taenia solium* metacestode excretion-secretion proteins. *Proteomics* **2012**, *12*, 1860–1869. [CrossRef] [PubMed]
206. Fayer, R. Sarcocystis spp. in Human Infections. *Clin. Microbiol. Rev.* **2004**, *17*, 894–902. [CrossRef] [PubMed]
207. Fayer, R.; Heydorn, A.O.; Johnson, A.J.; Leek, R.G. Transmission of Sarcocystis suihominis from humans to swine to nonhuman primates (*Pan troglodytes*, *Macaca mulatta*, *Macaca irus*). *Z. Parasitenkd. Parasitol. Res.* **1979**, *59*, 15–20. [CrossRef] [PubMed]
208. Andrews, R.H.; Sithithaworn, P.; Petney, T.N. Opisthorchis viverrini: An underestimated parasite in world health. *Trends Parasitol.* **2008**, *24*, 497–501. [CrossRef]
209. Traub, R.J.; Macaranas, J.; Mungthin, M.; Leelayoova, S.; Cribb, T.; Murrell, K.D.; Thompson, R.C.A. A New PCR-Based Approach Indicates the Range of Clonorchis sinensis Now Extends to Central Thailand. *PLoS Negl. Trop. Dis.* **2009**, *3*, e367. [CrossRef]
210. Wykoff, D.E.; Harinasuta, C.; Juttijudata, P.; Winn, M.M. Opisthorchis viverrini in Thailand: The Life Cycle and Comparison with *O. felineus*. *J. Parasitol.* **1965**, *51*, 207. [CrossRef]
211. Sripa, B.; Kaewkes, S.; Sithithaworn, P.; Mairiang, E.; Laha, T.; Smout, M.; Pairojkul, C.; Bhudhisawasdi, V.; Tesana, S.; Thinkamrop, B.; et al. Liver Fluke Induces Cholangiocarcinoma. *PLoS Med.* **2007**, *4*, e201. [CrossRef]
212. Prasopdee, S.; Thitapakorn, V.; Sathavornmanee, T.; Tesana, S. A comprehensive review of omics and host-parasite interplays studies, towards control of *Opisthorchis viverrini* infection for prevention of cholangiocarcinoma. *Acta Trop.* **2019**, *196*, 76–82. [CrossRef]
213. Boonmee, S.; Imtawil, K.; Wongkham, C.; Wongkham, S. Comparative proteomic analysis of juvenile and adult liver fluke, *Opisthorchis viverrini*. *Acta Trop.* **2003**, *88*, 233–238. [CrossRef]
214. Mulvenna, J.; Sripa, B.; Brindley, P.J.; Gorman, J.; Jones, M.K.; Colgrave, M.L.; Jones, A.; Nawaratna, S.; Laha, T.; Suttiprapa, S.; et al. The secreted and surface proteomes of the adult stage of the carcinogenic human liver fluke *Opisthorchis viverrini*. *Proteomics* **2010**, *10*, 1063–1078. [CrossRef] [PubMed]
215. Prasopdee, S.; Tesana, S.; Cantacessi, C.; Laha, T.; Mulvenna, J.; Grams, R.; Loukas, A.; Sotillo, J. Proteomic profile of *Bithynia siamensis* goniomphalos snails upon infection with the carcinogenic liver fluke *Opisthorchis viverrini*. *J. Proteom.* **2015**, *113*, 281–291. [CrossRef] [PubMed]
216. Suwannatrai, K.; Suwannatrai, A.; Tabsripair, P.; Welbat, J.U.; Tangkawattana, S.; Cantacessi, C.; Mulvenna, J.; Tesana, S.; Loukas, A.; Sotillo, J. Differential Protein Expression in the Hemolymph of *Bithynia siamensis* goniomphalos Infected with *Opisthorchis viverrini*. *PLoS Negl. Trop. Dis.* **2016**, *10*, e0005104. [CrossRef] [PubMed]
217. Wang, Q.P.; Lai, D.H.; Zhu, X.Q.; Chen, X.G.; Lun, Z.R. Human angiostrongyliasis. *Lancet Infect. Dis.* **2008**, *8*, 621–630. [CrossRef]
218. Martins, Y.C.; Tanowitz, H.B.; Kazacos, K.R. Central nervous system manifestations of *Angiostrongylus cantonensis* infection. *Acta Trop.* **2015**, *141PA*, 46–53. [CrossRef]
219. Sawanyawisuth, K.; Kitthaweesin, K.; Limpawattana, P.; Intapan, P.M.; Tiamkao, S.; Jitpimolmard, S.; Chotmongkol, V. Intraocular angiostrongyliasis: Clinical findings, treatments and outcomes. *Trans. R. Soc. Trop. Med. Hyg.* **2007**, *101*, 497–501. [CrossRef]
220. Mattis, A.; Mowatt, L.; Lue, A.; Lindo, J.; Vaughan, H. Ocular Angiostrongyliasis—First case report from Jamaica. *West Indian Med. J.* **2009**, *58*, 383–385.
221. Chen, K.Y.; Cheng, C.J.; Yen, C.M.; Tang, P.; Wang, L.C. Comparative studies on the proteomic expression patterns in the third- and fifth-stage larvae of *Angiostrongylus cantonensis*. *Parasitol. Res.* **2014**, *113*, 3591–3600. [CrossRef]
222. Huang, H.C.; Yao, L.L.; Song, Z.M.; Li, X.P.; Hua, Q.Q.; Li, Q.; Pan, C.W.; Xia, C.M. Development-Specific Differences in the Proteomics of *Angiostrongylus cantonensis*. *PLoS ONE* **2013**, *8*, e76982. [CrossRef]

223. Chen, K.Y.; Lu, P.J.; Cheng, C.J.; Jhan, K.Y.; Yeh, S.C.; Wang, L.C. Proteomic analysis of excretory-secretory products from the young adults of *Angiostrongylus cantonensis*. *Mem. Inst. Oswaldo Cruz* **2019**, *114*, e180556. [CrossRef]
224. Mega, J.D.; Galdos-Cardenas, G.; Gilman, R.H. Tapeworm Infections. In *Hunter's Tropical Medicine and Emerging Infectious Disease*; Elsevier: Amsterdam, The Netherlands, 2013; pp. 895–902.
225. Scholz, T.; Garcia, H.H.; Kuchta, R.; Wicht, B. Update on the Human Broad Tapeworm (Genus Diphyllobothrium), Including Clinical Relevance. *Clin. Microbiol. Rev.* **2009**, *22*, 146–160. [CrossRef] [PubMed]
226. Blair, D.; Agatsuma, T.; Wang, W. Paragonimiasis. In *Food-Borne Parasitic Zoonoses Fish and Plant-Borne Parasites*; Springer: New York, NY, USA, 2007; pp. 117–150.
227. Lee, E.G.; Na, B.K.; Bae, Y.A.; Kim, S.H.; Je, E.Y.; Ju, J.W.; Cho, S.H.; Kim, T.S.; Kang, S.Y.; Cho, S.Y.; et al. Identification of immunodominant excretory–secretory cysteine proteases of adult Paragonimus westermani by proteome analysis. *Proteomics* **2006**, *6*, 1290–1300. [CrossRef] [PubMed]
228. Park, H.; Kim, S.I.; Hong, K.M.; Kim, M.J.; Shin, C.H.; Ryu, J.S.; Min, D.Y.; Lee, J.B.; Hwang, U.W. Characterization and classification of five cysteine proteinases expressed by *Paragonimus westermani* adult worm. *Exp. Parasitol.* **2002**, *102*, 143–149. [CrossRef]
229. Chai, J.Y. *Human Intestinal Flukes*; Springer: Dordrecht, The Netherlands, 2019; ISBN 978-94-024-1702-9.
230. Fried, B.; Graczyk, T.K.; Tamang, L. Food-borne intestinal trematodiases in humans. *Parasitol. Res.* **2004**, *93*, 159–170. [CrossRef]
231. Jadhav, S.R.; Shah, R.M.; Karpe, A.V.; Morrison, P.D.; Kouremenos, K.; Beale, D.J.; Palombo, E.A. Detection of foodborne pathogens using proteomics and metabolomics-based approaches. *Front. Microbiol.* **2018**, *9*, 1–13. [CrossRef]
232. Rousseau, A.; La Carbona, S.; Dumètre, A.; Robertson, L.J.; Gargala, G.; Escotte-Binet, S.; Favennec, L.; Villena, I.; Gérard, C.; Aubert, D. Assessing viability and infectivity of foodborne and waterborne stages (cysts/oocysts) of *Giardia duodenalis*, *Cryptosporidium* spp., and *Toxoplasma gondii*: A review of methods. *Parasite* **2018**, *25*, 14. [CrossRef]
233. Gamble, H.R.; Murrell, K.D. Detection of parasites in food. *Parasitology* **1999**, *117*, 97–111. [CrossRef]
234. Aslam, B.; Basit, M.; Nisar, M.A.; Khurshid, M.; Rasool, M.H. Proteomics: Technologies and their applications. *J. Chromatogr. Sci.* **2017**, *55*, 182–196. [CrossRef]
235. Jagadeesh, D.S.; Kannegundla, U.; Reddy, R.K. Application of proteomic tools in food quality and safety. *Adv. Anim. Vet. Sci.* **2017**, *5*, 213–225. [CrossRef]
236. Bassols, A.; Turk, R.; Roncada, P. A Proteomics Perspective: From Animal Welfare to Food Safety. *Curr. Protein Pept. Sci.* **2014**, *15*, 156–168. [CrossRef]
237. Borràs, E.; Sabidó, E. What is targeted proteomics? A concise revision of targeted acquisition and targeted data analysis in mass spectrometry. *Proteomics* **2017**, *17*, 1700180. [CrossRef] [PubMed]
238. Papadopoulos, M.C.; Abel, P.M.; Agranoff, D.; Stich, A.; Tarelli, E.; Bell, B.A.; Planche, T.; Loosemore, A.; Saadoun, S.; Wilkins, P.; et al. A novel and accurate diagnostic test for human African trypanosomiasis. *Lancet* **2004**, *363*, 1358–1363. [CrossRef]
239. Rioux, M.C.; Carmona, C.; Acosta, D.; Ward, B.; Ndao, M.; Gibbs, B.F.; Bennett, H.P.; Spithill, T.W. Discovery and validation of serum biomarkers expressed over the first twelve weeks of *Fasciola hepatica* infection in sheep. *Int. J. Parasitol.* **2008**, *38*, 123–136. [CrossRef] [PubMed]
240. Deckers, N.; Dorny, P.; Kanobana, K.; Vercruysse, J.; Gonzalez, A.E.; Ward, B.; Ndao, M. Use of ProteinChip technology for identifying biomarkers of parasitic diseases: The example of porcine cysticercosis (*Taenia solium*). *Exp. Parasitol.* **2008**, *120*, 320–329. [CrossRef] [PubMed]
241. Santamaria, C.; Chatelain, E.; Jackson, Y.; Miao, Q.; Ward, B.J.; Chappuis, F.; Ndao, M. Serum biomarkers predictive of cure in Chagas disease patients after nifurtimox treatment. *BMC Infect. Dis.* **2014**, *14*, 302. [CrossRef]
242. Sánchez-Ovejero, C.; Benito-Lopez, F.; Díez, P.; Casulli, A.; Siles-Lucas, M.; Fuentes, M.; Manzano-Román, R. Sensing parasites: Proteomic and advanced bio-detection alternatives. *J. Proteom.* **2016**, *136*, 145–156. [CrossRef]
243. Newell, D.G.; Koopmans, M.; Verhoef, L.; Duizer, E.; Aidara-Kane, A.; Sprong, H.; Opsteegh, M.; Langelaar, M.; Threfall, J.; Scheutz, F.; et al. Food-borne diseases—The challenges of 20 years ago still persist while new ones continue to emerge. *Int. J. Food Microbiol.* **2010**, *139*, S3–S15. [CrossRef]

244. Klimpel, S.; Palm, H.W. Anisakid Nematode (Ascaridoidea) Life Cycles and Distribution: Increasing Zoonotic Potential in the Time of Climate Change? In *Progress in Parasitology. Parasitology Research Monographs*; Mehlhorn, H., Ed.; Springer: Berlin/Heidelberg, Germany, 2011; pp. 201–222. ISBN 978-3-642-21395-3.
245. Audicana, M.T.; Kennedy, M.W. Anisakis simplex: From Obscure Infectious Worm to Inducer of Immune Hypersensitivity. *Clin. Microbiol. Rev.* **2008**, *21*, 360–379. [CrossRef]
246. Aibinu, I.E.; Smooker, P.M.; Lopata, A.L. Anisakis Nematodes in Fish and Shellfish- from infection to allergies. *Int. J. Parasitol. Parasites Wildl.* **2019**, *9*, 384–393. [CrossRef]
247. Audicana, M.T.; Ansotegui, I.J.; de Corres, L.F.; Kennedy, M.W. Anisakis simplex: Dangerous—Dead and alive? *Trends Parasitol.* **2002**, *18*, 20–25. [CrossRef]
248. D'Amelio, S.; Lombardo, F.; Pizzarelli, A.; Bellini, I.; Cavallero, S. Advances in Omic Studies Drive Discoveries in the Biology of Anisakid Nematodes. *Genes* **2020**, *11*, 801. [CrossRef] [PubMed]
249. Fæste, C.K.; Jonscher, K.R.; Dooper, M.M.W.B.; Egge-Jacobsen, W.; Moen, A.; Daschner, A.; Egaas, E.; Christians, U. Characterisation of potential novel allergens in the fish parasite Anisakis simplex. *EuPA Open Proteom.* **2014**, *4*, 140–155. [CrossRef] [PubMed]
250. Polak, I.; Łopieńska-Biernat, E.; Stryiński, R.; Mateos, J.; Carrera, M. Comparative proteomics analysis of Anisakis simplex s.s.—Evaluation of the response of invasive larvae to ivermectin. *Genes* **2020**, *11*, 11060710. [CrossRef] [PubMed]
251. Rodríguez-Mahillo, A.I.; González-Muñoz, M.; de las Heras, C.; Tejada, M.; Moneo, I. Quantification of Anisakis simplex Allergens in Fresh, Long-Term Frozen, and Cooked Fish Muscle. *Foodborne Pathog. Dis.* **2010**, *7*, 967–973. [CrossRef] [PubMed]
252. Fæste, C.K.; Plassen, C.; Løvberg, K.E.; Moen, A.; Egaas, E. Detection of Proteins from the Fish Parasite Anisakis simplex in Norwegian Farmed Salmon and Processed Fish Products. *Food Anal. Methods* **2015**, *8*, 1390–1402. [CrossRef]
253. Fæste, C.K.; Moen, A.; Schniedewind, B.; Haug Anonsen, J.; Klawitter, J.; Christians, U. Development of liquid chromatography-tandem mass spectrometry methods for the quantitation of Anisakis simplex proteins in fish. *J. Chromatogr. A* **2016**, *1432*, 58–72. [CrossRef]
254. Carrera, M.; Gallardo, J.M.; Pascual, S.; González, Á.F.; Medina, I. Protein biomarker discovery and fast monitoring for the identification and detection of Anisakids by parallel reaction monitoring (PRM) mass spectrometry. *J. Proteom.* **2016**, *142*, 130–137. [CrossRef]

© 2020 by the authors. Licensee MDPI, Basel, Switzerland. This article is an open access article distributed under the terms and conditions of the Creative Commons Attribution (CC BY) license (http://creativecommons.org/licenses/by/4.0/).

Review

Proteomic Strategies to Evaluate the Impact of Farming Conditions on Food Quality and Safety in Aquaculture Products

Mónica Carrera [1,*], Carmen Piñeiro [2] and Iciar Martinez [3,4]

1. Food Technology Department, Institute of Marine Research (IIM), Spanish National Research Council (CSIC), 36208 Vigo, Pontevedra, Spain
2. Scientific Instrumentation and Quality Service (SICIM), Institute of Marine Research (IIM), Spanish National Research Council (CSIC), 36208 Vigo, Pontevedra, Spain; cpineiro@iim.csic.es
3. Research Centre for Experimental Marine Biology and Biotechnology—Plentzia Marine Station (PiE), University of the Basque Country UPV/EHU, 48620 Plentzia, Spain; iciar.martinez@ehu.eus
4. IKERBASQUE Basque Foundation for Science, 48013 Bilbao, Spain
* Correspondence: mcarrera@iim.csic.es; Tel.: +34-986-231930; Fax: +34-986-292762

Received: 9 July 2020; Accepted: 23 July 2020; Published: 4 August 2020

Abstract: This review presents the primary applications of various proteomic strategies to evaluate the impact of farming conditions on food quality and safety in aquaculture products. Aquaculture is a quickly growing sector that represents 47% of total fish production. Food quality, dietary management, fish welfare, the stress response, food safety, and antibiotic resistance, which are covered by this review, are among the primary topics in which proteomic techniques and strategies are being successfully applied. The review concludes by outlining future directions and potential perspectives.

Keywords: proteomics; discovery; target; aquaculture; mass spectrometry; dietary management; fish welfare; stress; food safety; antibiotic resistance

1. Introduction to Proteomics in Aquaculture

Aquaculture is the breeding of aquatic organisms under controlled conditions, involving both marine and freshwater fish along with algae, crustaceans, and mollusks. According to the Food and Agriculture Organization of the United Nations (FAO), this food sector represents a significant source of nutrients for the human diet and produces approximately 97.2 million tons of fish annually, which represents 47% of the global fish production [1]. With nine billion people expected to be living on the planet by 2050, maintaining the current level of fish consumption (9.0–20.2 kg annually per capita) is a challenging task [1]. Aquaculture, the most rapidly growing food-producing sector in the world, offers an excellent source of high value food and is expected to significantly contribute to meeting this demand for fish products.

The globalization of aquaculture markets presents important nutritional and economic benefits but also poses potential risks for food safety, such as the fraudulent substitution of fish species and the presence of food microorganisms, viruses, parasites, and vectors of their corresponding foodborne diseases [2]. Moreover, fish and seafood are easily spoiled, resulting in a fast loss of food quality due to the presence of fish microbiota, the elevated amount of unsaturated fatty acids and the abundance of proteases. Improving aquaculture practices to offer products of optimal quality and to reduce chronic stress throughout improved farming conditions to maintain fish welfare are two major questions in aquaculture research. Consumer awareness and a rising demand for aquaculture products have motivated scientists to develop procedures to enhance productivity and to improve the quality and safety of these foodstuffs. In this context, proteomics has been established as a powerful methodology for the evaluation of quality and safety in aquaculture products [3–5].

Proteomics is the high-throughput analysis of the proteins of a specific biological sample [6]. Proteomics involves the identification, localization and quantification of proteins as well as the analysis of protein modifications and the elucidation of protein-protein networks [7]. Among proteomic analytical techniques, mass spectrometry (MS) is recognized as an indispensable instrument to precisely analyze a large number of proteins from complex samples in the majority of food proteomics studies [8,9]. Additionally, the computational analysis of MS data has improved the discriminatory power of proteomics techniques, making them effective methodologies for the global analysis of proteins and peptides [10]. Thus, the latest advances in proteomics and bioinformatics approaches have turned them into useful tools to develop promising strategies for food science investigations [11,12]. Within that framework, the present review summarizes some highly relevant applications of proteomics to evaluate the impact of farming conditions in fish wellbeing as well as the quality and safety of aquaculture products.

2. Workflow of Proteomics: Discovery and Targeted Proteomics

Proteomics has the potential to provide information useful to improve the production, welfare, health, nutritional value and wholesomeness of farmed fish. Figure 1 shows the classical proteomics approaches, i.e., discovery and targeted proteomics, with their corresponding workflows.

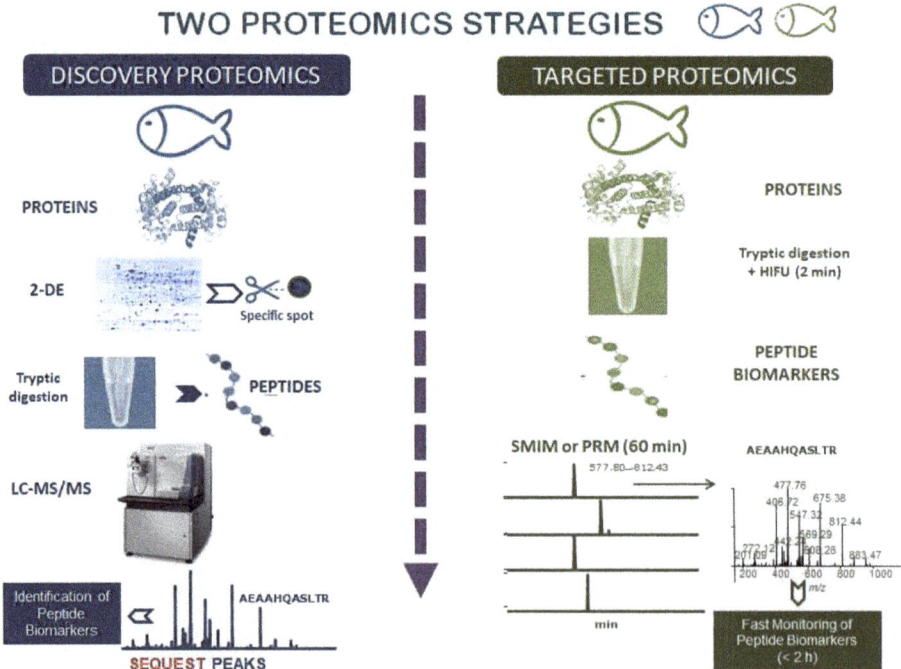

Figure 1. Workflow of proteomics: discovery and targeted proteomics.

Discovery proteomics aims at identifying biological markers in a given proteome, frequently employing a bottom-up approach, in which the proteins of the sample are separated, proteolyzed with enzymes such as trypsin or Glu-C and the peptides obtained are subsequently analyzed by tandem mass spectrometry (MS/MS). Two-dimensional gel electrophoresis (2-DE) has traditionally been the technique selected for the separation of proteins samples [13]. This gel-based procedure is the most suitable approach for species whose protein sequences are not yet known, which includes many fish. In these cases, identification is performed by comparison of the MS/MS spectra of the peptides obtained with orthologous protein sequences from related species or by de novo MS/MS sequencing [14]. The 2-DE gels themselves can be analyzed by programs such as Progenesis and PDQuest.

In gel-free approaches, also known as shotgun proteomics, the proteins are directly digested in the extract with a selected enzyme, and the obtained mixture of peptides is subsequently analyzed by liquid chromatography (LC) coupled to tandem mass spectrometry (LC-MS/MS) [15,16]. It is possible to perform multidimensional LC separations, combining, for example, strong anion/cation exchange chromatography (SA/CX) and reverse phase (RP) chromatography [17]. Database searching programs, like SEQUEST, X! Tandem, or Mascot [18,19], allow the tentative identification of presumed peptide sequences based on the obtained fragmentation spectra, and additional software programs, such as Percolator are used to validate the identification [20]. When the protein is not present in the database, then the peptides must be sequenced de novo [21], either manually or using programs such as PEAKS and DeNovoX [22,23]. This approach has been successfully used in the de novo sequencing of some fish allergens, such as parvalbumins and shrimp arginine kinases [14,24,25]. When protein quantification is deemed necessary, the methods of choice include metabolic stable isotope labeling (such as stable isotope labeling by/with amino acids in cell culture, SILAC) [26]; isotope tagging by chemical reaction, such as isobaric tags for relative and absolute quantitation (iTRAQ), tandem mass tag (TMT) and difference gel electrophoresis (DIGE) [27–29]; stable isotope incorporation via enzyme reaction (i.e., ^{18}O) [30]; and label-free quantification (i.e., measuring the intensity of the peptides at the MS level) [31]. After matching the obtained peptides and proteins by alignment software programs like BLAST (https://blast.ncbi.nlm.nih.gov/), it is possible to select relevant peptide biomarkers to be used in the subsequent phase namely, targeted proteomics.

Targeted proteomics refers to the monitoring of the relevant peptide biomarkers and it has become a recognized methodology to detect selected proteins with significant accuracy, reproducibility, and sensitivity [32]. In targeted proteomics, the MS analyzer is focused on detecting only the peptide/s chosen by selected/multiple-reaction monitoring (SRM/MRM) [33]. Monitoring appropriate transitions (evens of precursor and fragment ions m/z), represents a common analysis for detecting and identifying peptide biomarkers. These techniques are selective, sensitive, highly reproducible, with a high dynamic range and an excellent signal-to-noise (S/N) ratio [34]. SRM/MRM modes are usually performed on triple quadrupole (QQQ) instruments. This method possesses a highly sensitive scanning procedure but its optimization for a final SRM/MRM analysis is very time-consuming and, most importantly, this scanning mode does not produce entire MS/MS spectra. Since the spectrum of a peptide is critical to verify its sequence, new procedures are being used to obtain entire structural information; for instance, SRM-triggered MS/MS using hybrid quadrupole-ion trap (Q-IT) mass spectrometers, selected MS/MS ion monitoring (SMIM), parallel reaction monitoring (PRM) in IT or high-resolution Q-Orbitrap instruments are alternative targeted modes that enable the monitoring of precise peptides [35–37]. The development of targeted data independent analysis (DIA), conducted on a sequential windowed acquisition of all theoretical fragment ion spectra (SWATH-MS) [38], can identify and quantify thousands of proteins without the prerequisite of specifying a group of proteins prior to analysis. Stable-isotope dilution, ^{13}C- or ^{15}N-labeled absolute quantification peptide standards (AQUA) or concatemer of standard peptides (QCAT) can also be introduced to the sample as internal standards for absolute quantification of the proteins [39]. Programs such as SRMCollider and Skyline are accessible for the analysis of different targeted proteomic modes [40,41]. The following sections will show the application of the scanning mode for the follow up of peptide biomarkers identified in the discovery phase to assess the impact of farming conditions on food quality and safety of farmed fish.

3. Application of Proteomics to Evaluate the Farming Conditions on Food Quality and Safety in Aquaculture Products

Fish farming environments and conditions are very different from the conditions in which fish live in nature. Farmed fish, for instance, do not need to actively swim to catch their prey or escape predators; therefore, they exercise less, which impacts on their muscle growth and phenotype. The heavily processed feed consumed by farmed fish differs considerably from their natural diet, which affects muscle metabolism and biochemical composition. In addition, the farming conditions in aquaculture may not be optimized regarding stocking densities, incidence of parasites and diseases, and establishment of hierarchies due to competition for space or feed, all of which have consequences for the wellbeing and development of abnormal behavior. All these variables exert a strong influence on the yield, quality and wholesomeness of farmed seafood. Moreover, development of analytical methods to ensure that fish was farmed minimizing stressful factors is also of high relevance to satisfy consumer demands and labeling on the welfare of fish to be used for food. To investigate all these topics, powerful proteomic methodologies (discovery and targeted proteomics) may have a considerable impact on the understanding of current aquaculture practices in several major areas: (i) dietary management, (ii) fish welfare and response to stress, (iii) food safety, and (iv) antibiotic resistance (Figure 2).

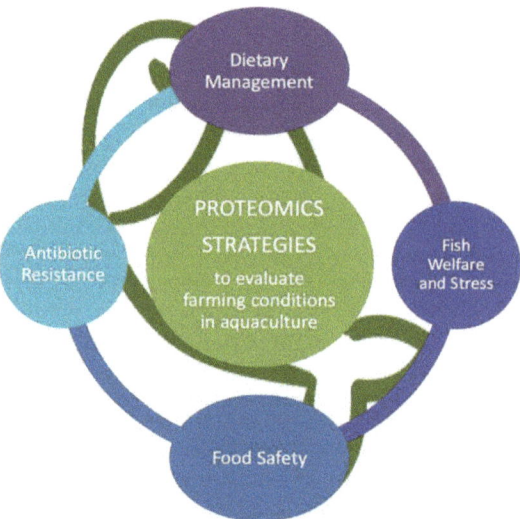

Figure 2. Summary of the main applications of proteomics techniques to evaluate the farming conditions in aquaculture reviewed in this publication.

3.1. Dietary Management in Aquaculture

Dietary management in aquaculture attempts to improve growth performance, health and immune status in living aquaculture organisms. Numerous works have shown that the composition of the feed influences the composition of fish fillet. Initial works using 2-DE revealed differences between the proteomes of skeletal muscle samples from wild and farmed fish. Carpene et al. [42], found differences by 2-DE in the level of abundance of the fast skeletal myosin light chain type 3 which seemed to be more abundant in wild than in farmed fish and, surprisingly, was also present in the red muscle of farmed, but not of wild, fish [42]. The 2-DE protein pattern of skeletal muscle of cod excised within 5 h of death revealed the presence of spots in the ranges of molecular weight between 35 and 45 kDa and between 50 and 100 kDa in the muscles of the farmed fish that were not present in the wild cod [43]. The authors attributed the differences in the proteome to differences during cultivation that may have affected not only the make-up of the muscle in vivo, but also the postmortem muscle conditions (for example, pH) and the abundance and regulation of proteases relevant in postmortem muscle tenderization. A large scale study by Chiozzi and coworkers [44]. comparing the proteome of wild and farmed European seabass with that of wild specimens from the same area in the Mediterranean by label-free multidimensional shotgun proteomics to identify relationships between farming conditions and quality and safety of the fish, confirmed muscle atrophy in farmed fish [44]. The most abundant upregulated proteins in farmed sea bass were some structural proteins and proteins involved in binding and catalytic activities, while the main downregulated proteins also involved catalytic activities and binding.

Optimization of fish diets has been a priority in the aquaculture sector for many years [45]. One ingredient whose incorporation in feed seems to improve growth performance and the humoral immune response of some fish species is β-glucan [46]. Feeding β-glucan to rainbow trout induced an increase in the amounts of tropomyosin isoforms and it lowered those of myosin light and heavy chain isoforms in the proteome of the fillet in treated trout [47]. Evaluation of the effects of partial substitution of fish meal by plant proteins on the fish proteome in different tissues has also been the target of several studies. Thus, partial substitution of fish meal with soybean meal caused an increase in the amount of enzymes involved in protein catabolism and turnover in the liver of rainbow trout [48] and it affected the proteome of gut mucosa in gilthead bream [49].

Reduction of the use of fish meal and oil in aquaculture is a priority [45], which has led to investigate the effects of novel diets on fish physiology where some marine ingredients were substituted by vegetable protein and oils [50]. The inclusion of vegetable oil feed induced a specific response in the intestinal proteome in salmonids, indicating a defense against oxidative cellular stress [51]. Significantly downregulated proteins were those related to oxidative stress and motility, including the myosin light chains, peroxiredoxin-1 and hemopexin-like protein.

Using analytical techniques based on microfluidic electrophoresis and sequencing, some authors have consistently confirmed differences in the protein abundance and/or regulation in fish muscles depending on the production method. Monti et al. [52], using SDS-PAGE, MALDI and ESI MS/MS, and CE for protein identification and relative quantification, showed that the enzymes involved in the metabolism of carbohydrates were upregulated in farmed sea bass muscle (i.e., glyceraldehyde-3-phosphate dehydrogenase and aldolase), while creatine kinase, nuclease diphosphate kinase B and parvalbumin were downregulated, displaying the expected proteome pattern of muscle in farmed fish [52]. Additionally, new protein sources, such as insect meal, have been characterized for aquafeeds by direct comparison through LC-MS/MS analysis [53].

3.2. Fish Welfare and Stress Response in Aquaculture

The effects of stress on growth have been studied extensively in animal production and aquaculture [54]. Different chronic stress conditions, such as confinement, overcrowding, repetitive handling, deficient water and diet and hypoxia, affect the welfare and stress response of aquaculture organisms and several proteomic studies have identified robust protein signatures for chronic stress in fish [55]. Elevated cortisol due to long-term stress conditions has a strong impact on the entire organism and is directly linked to the inhibition of muscle growth by inhibiting protein synthesis and increasing protein catabolism to obtain energy from amino acids [56]. The proteome of fish farmed under these stressful conditions displays an increase in the amount of enzymes related to protein catabolism, and a decrease in the amount of the structural proteins, with the latter being degraded to provide energy. Stress-related depletion of the required energy for muscle growth has been shown to lead to muscle atrophy [57].

Farming itself affects the levels of stress the fish suffer, their growth, and the biochemical composition of different tissues and organs, including the liver, brain, and muscle. For instance, exercising in salmonids lowers the levels of aggression and the building up of hierarchies, leading to increased fish welfare and growth [58–60]. It is reasonable to assume that these observations would apply to any species with similar behavioral characteristics, i.e., active swimmers with shoaling behavior and schooling responses [61]. Interestingly, while muscle atrophy is provoked by not using the muscle, its use induces both muscle growth and a type of muscle damage due to the need to develop and grow both the activated muscle satellite cells (to regenerate the lesion) and the existing myofibrils that need to increase their volume, i.e., inducing both muscle hyperplasia and hypertrophy [62]. Thus, while normal muscle growth in adult fish will be accomplished by hyperplasia and, mostly, by hypertrophy [63–66], muscle regeneration and subsequent growth, as observed in exercising fish [62], recapitulates embryonic myogenesis through the activation of satellite cells and the consequent larger contribution from hyperplasia followed by hypertrophy of the muscle fibers [64,67]. This process involves alterations in protein abundance and muscle metabolism to achieve muscle growth [68] and should ultimately lead to an increase in the yield of fillet and to improved welfare of the fish.

One study on sea bream, however, was not able to show consistent differences between the 2-DE patterns of muscle from two natural repopulation lagoons and those of fish from four offshore mariculture plants in Italy by 2-DE, MALDI-MS and LC-ESI-Q-TOF-MS [69]. The similarity between the proteomes of farmed and wild fish would indicate the suitability of the farming conditions and locations. The authors did, however, find significant individual differences in the relative expression of parvalbumin isoforms and of spots corresponding to the myosin-binding protein H (MyBP-H) isoelectric series with no apparent relationship to the length of the fish, its production method or geographical location [69]. Muscle protein patterns obtained by 2-DE analysis showed variations attributed to different factors; for example, acclimation to higher water temperature significantly increased the amount of the warm temperature acclimation-related protein-65 isoforms, and the ratio of structural proteins vs. glycolytic enzymes increased as fish grew larger [69]. This work is particularly interesting because it shows that it is possible to achieve offshore-farmed gilthead sea breams of commercial size whose protein expression profile is comparable to that of wild fish [69].

Discovery proteomics has been applied to improve our understanding of the mechanisms implicated in skeletal deformities [70]. Analysis of how preslaughter stress affects the postmortem processes in gilthead seabream muscle was performed by 2-DE and MALDI-TOF-TOF MS [70]. Moreover, 2-DE followed by LC-MS/MS of grass carp gills uncovered alterations in the metabolic pathways after hypoxic stress [71], some of which were involved in energy generation, metabolic, immunity and oxidative processes and proteolytic activities. It must be emphasized that the improvement of farming practices, leading to minimizing chronic stress and preserving fish welfare not only is one of the primary challenges for fish farmers, it is also a demand by European consumers.

The Pacific geoduck clam is one of the species whose farming is seeing a successful bloom and, consequently, a species under study to improve its production [72]. Proteomic studies on how geoduck production may be impacted by conditions susceptible of being modified by ocean acidification, such as pH and temperature were performed by Spencer et al. [73]. The results showed that the amounts of heat shock protein 90-α, puromycin-sensitive aminopeptidase and tri-functional-enzyme β-subunit as well as shell growth, kept a negative correlation with the average temperature and a positive one with the amount of dissolved oxygen. That indicates that geoducks may be more resistant to acidification under natural conditions and more susceptible to variations in the concentration of dissolved oxygen and the temperature of the water.

3.3. Food Safety in Aquaculture

Aquaculture has the capacity to provide food for millions of people over the world, however inappropriate facility management may severely damage aquatic ecosystems and affect health risks to consumers through contamination with environmental or human-made hazards. Moreover, several pathogenic microorganisms can be found in the aquatic environment with the potential to negatively affect, not only aquatic life, but also human health. In the last decade, the risk of dissemination of infectious or toxic agents and the occurrence of disease outbreaks has risen mainly due to increases in the following factors: (i) intake of raw or scarcely processed seafood; (ii) international trade of aquaculture products; (iii) suboptimal monitoring methodologies; (iv) alterations in ecological stability;

and (v) contamination and climatic change [74]. Food safety in the aquaculture sector is of crucial relevance to avoid health hazards that can be biotic (bacteria, allergies, parasites, virus or harmful algae blooms) and abiotic (aromatic hydrocarbons, dioxins, heavy metals, plastics) [75]. To control and minimize the presence of hazards, the FAO implemented a code of practice for aquaculture products [1], where the handling of fish is presented according the requirements of HACCP.

To make the reading of this review more comprehensible, the authors have decided to divide the hazards to which consumers of aquaculture products are exposed into two general groups: biotic hazards and abiotic hazards.

3.3.1. Food Safety in Aquaculture: Biotic Hazards

Foodborne poisonings are a relevant cause of mortality and morbidity which result from drinking water or eating food contaminated with such pathogens as viruses, bacteria and parasites, and their toxins.

Regarding biotic hazards, there are two main groups of bacteria that affect food products of aquaculture: those naturally present in its habitat (*Aeromonas* spp., *Clostribuim botulinum*, *Listeria monocytogenes*, *Vibrio cholerae,* and *Vibrio parahaemolyticus*) and those derived from environmental contamination (Enterobacteriaceae, *Escherichia coli*, and *Salmonella* spp.) [76]. Additionally, *Staphylococcus aureus* can infect aquaculture species during management due to inadequate hygiene conducts of operators in the processing factories [77,78].

Proteomics has been applied to the detection and identification of bacterial species from aquaculture products both after their direct detection in fish products, or after their isolation and growing in different culture media. For instance, MALDI-TOF-MS techniques enabled us to achieve mass spectral fingerprints of *Vibrio* spp., a Gram-negative bacteria causative of gastrointestinal diseases in humans after ingestion of poorly cooked infected seafood, such as seabream and mollusks [79,80]. The exhaustive proteome and transcriptome data analysis obtained by the study of Li and coworkers by iTRAQ coupled to MRM provided some critical protein signatures for the study of the regulatory mechanisms of the intestinal mucosal immunity in grass carp (*Ctenopharyngodon idella*) against *Vibrio mimicus* [81]. Vaccination altered the regulation of 5339 genes and of 1173 proteins in the grass carp intestines. The conclusions of the study suggest that the integration of the five activated immune-related pathways is relevant to the improved immune response of the intestinal mucosal in immunized carp. MALDI-TOF MS analyses have been performed to obtain available reference spectral libraries for diverse bacterial strains isolated from seafoods [82]. Recently, the open MALDI Biotyper library (Bruker MALDI Biotyper) allowed for the precise identification of 75 pathogenic bacterial isolates [83].

Targeted proteomics applications in the field of pathogenic bacteria have increased substantially in recent years. For example, both SRM and PRM have produced sensitive quantitative results about the proteins associated with bacterial infection, particularly in the fields of clinical diagnosis and antibiotic resistance [84]. Thus, iTRAQ-based quantitative proteomics followed by MRM studies were used to contrast the differentially regulated proteins of *Aeromonas veronii*. This bacterium, a Gram-negative virulent pathogen associated with infections in freshwater fish species and mammals, is capable of adhering to biotic and abiotic surfaces surrounded by the extracellular matrix produced by the resident microorganisms. The study, which used an in vitro biofilm model [85], showed that the upregulated TonB protein increased the nutrient absorption capacity, and the enolase gene was involved in the regulation of multiple pathways, leading to enhancement of the bacteria's ability to undergo invasion and metastasis. These changes may be the principal cause for the capability of *A. veronii* to create biofilms and its increased dissemination.

Protozoan parasites, such as *Ichthyophthirius multifiliis*, cause important economic losses to the aquaculture sector. Upon exposure to the parasite, LC-ESI-MS/MS revealed the differential regulation of some immune-related signal transduction proteins in the skin mucus of common carp [86]. Multiple lectins and several serpins with protease inhibitor activity were likely implicated in lectin pathway activation and regulation of proteolysis, indicating that these proteins support the carp innate immune system and the preventive characteristics of the skin mucus.

Virus infections can decimate the production in fish and shrimp farms. Among the latter, white spot syndrome virus is currently one of the most serious global hazards. A protein interactomics map for the white spot syndrome virus has been produced by means of a co-immunoprecipitation assay (co-IP) from a yeast two-hybrid approach [87].

Harmful algal blooms (HABs) generate shellfish poisoning toxins that affect aquaculture, particularly mussel farming. Gel-based proteomic approaches were used to distinguish and identify nontoxic dinoflagellates from the toxic dinoflagellate *Alexandrium tamarense* [88]. An alternative approach consists of the generation and application of monoclonal antibodies directed against intracellular antigens of the toxic dinoflagellate *Alexandrium minutum* as described by Carrera et al. (2010) [89]. Recent proteomic studies have contributed to enlarge the volume of data in sequence databases suitable to identify how the proteomes of HABs are modulated by physiological parameters and in response to changes in the environment, such as climate change [90]. In that regard, Piñeiro et al. (2010) [91] reviewed the application of proteomics methods to study the effects of climate change on the quality and safety of wild and cultivated seafood products.

Proteomic studies and systems biology analysis of allergenic proteins have also been critical determinants for the evaluation of the quality and safety of wild and cultivated fish and crustacean food products [92]. The major allergen identified in fish is β-parvalbumin. A rapid strategy for the detection of fish β-parvalbumin in fish products was performed by targeted proteomics using SMIM [93]. On the other hand, tropomyosin is the major allergen in shrimp and mollusks. Proteomic profiling of the allergen tropomyosin was performed to obtain the full amino acid sequence in a Q-TOF instrument [94]. Recently, the impact of EDTA-enriched diets on farmed fish allergenicity was studied by 2-DE [95].

3.3.2. Food Safety in Aquaculture: Abiotic Hazards

Abiotic hazards in aquaculture have been extensively studied in mollusks exposed to contaminants in polluted areas. Discovery proteomics studies have been mainly performed for environmental assessment and marine pollution monitoring using the digestive glands of mussels *Mytilus galloprovincialis* by 2-DE and MS [96]. Bottom-up proteomics approaches on mussels exposed to fresh fuel and weathered fuel in a laboratory experiment that attempted to mimic the effects of the Prestige's oil spill were performed by 2-DE and MS [97]. Moreover, 2-DE and MALDI-TOF/TOF MS analyses have also been applied to the identification of differentially regulated proteins in the gonads of the oyster *Crassostrea angulata* after $HgCl_2$ contamination [98]. The first shotgun proteomics analysis of mussels after exposure to pharmaceutical environmental contaminants, such as propanolol, was performed by Campos et al. (2016) [99].

To assess complex field contamination, targeted proteomics using SRM methodologies was applied to the quantification of dozens of protein biomarkers in caged amphipods (*Gammarus fossarum*) after in situ exposure to several aquatic environments [100]. The work detected some of the previously identified and currently well-established protein biomarkers for amphipod crustaceans, such as the detoxification/antioxidant enzymes glutathione S-transferase, acetylcholinesterase, catalase, superoxide dismutase and some digestive enzymes [100].

Nanoparticle pollution is a recent issue of concern that has also been addressed by proteomic techniques. Thus, targeted proteomics has shown that the ionic form of Ag impacted on the growth of *Pseudomonas* spp. more strongly than did the nanoparticulate form of Ag in a bacterium isolated from waters in a region where fish is farmed for human consumption [101]. Possessing broad-spectrum antimicrobial properties, silver nanoparticles (AgNPs) are widely used in textiles and medical drugs. Approximately 20–130 tons of ionic silver (Ag^+) have been predicted to reach EU freshwaters annually, mostly due to leaching of ionic AgNPs from biocidal plastics and textiles. Proteomic analysis using SWATH-MS allowed the identification of 166 proteins affected by exposure to the nanoparticulate form of Ag, which also impacted on the growth of *Pseudomonas* spp. The form of Ag induced different adaptive responses in the metabolic, stress, and energetic pathways in *Pseudomonas* spp., and proteins affected were transmembrane transporters, chaperones, and proteins related to the metabolism of carbohydrates and proteins, indicating their potential value as biomarkers of the stress induced by Ag^+ and/or AgNPs. Among all the modified proteins, 59 had their content significantly changed by one or both forms of silver. In view of the evidences obtained in these studies, we believe that nanoparticle pollution should be considered an emergent hazard in waters with aquaculture production.

3.4. Antibiotic Resistance in Aquaculture

Antibiotics are natural and synthetic compounds that kill bacteria and have been heavily used and abused in aquaculture for over 50 years [102] not only to treat and prevent infections, but also to promote growth. Fortunately, success in the development of vaccines for the most relevant infections and in the implementation of vaccination programs has greatly reduced their use in some countries (e.g., Norway) although it is still a very serious problem in other countries and in the breeding of some species.

The abuse in antibiotic treatments has provoked the development and spreading of bacterial resistance and the appearance and expansion of multidrug-resistant strains in such a way that the aquatic environment has become an important reservoir of antibiotic resistance genes/proteins (ARG/Ps) and a route for their dissemination and potential transmission to human pathogens. Until now, five main mechanisms of antibiotic resistance have been accurately recorded due to the related development of resistance to drugs that (a) deteriorate enzymes, (b) bypass target pathways, (c) change antibiotic focus sites, (d) alter the penetrability of porins, and/or (e) trigger flow systems [103].

Proteomics techniques have the potential to significantly contribute to increase the knowledge about molecular mechanisms related to antibiotic resistance [104]. In the last five years, metagenomics and metaproteomics have been applied to identify correlations between the "resistome" (the antibiotic immunity genes/proteins) and the transmission of ARG/Ps from natural microflora to human pathogenic microorganisms, which could become a serious health issue. Conventional methodologies to evaluate water quality had been used to analyze marine sediments close to aquaculture farms, evidencing that the native "resistome" had been enriched by the use of antibiotics at the farming sites, although the findings were restricted to only a group of genes/proteins [5].

Proteomic studies on antibiotic resistance of fish- and shellfish-borne bacteria have largely been performed on *Aeromonas* spp., because this pathogen is responsible of hemorrhagic septicemia and hemolytic diseases in aquaculture, which cause large financial losses to farmers. Some of these studies are described below.

Tetracyclines are commonly used antibiotics comprising the monocyclines group, doxycycline, and chlortetracycline (CTC), and they are very efficient against both gram-positive and gram-negative microbes. In aquaculture, tetracycline resistant *Aeromonas hydrophila* (a notorious pathogen causing infections in many relevant wild and farmed species including carp, shellfish, grass carp and shrimp) has been confirmed by different proteomics analyses. Comparison between the fitness and acquired resistance to CTC in an *Aeromonas hydrophila* biofilm by TMT-labeling-based quantitative proteomics indicated an increase in translation-related ribosomal proteins in both cases and an increase in proteins involved in fatty acid biosynthesis only in biofilm fitness, while proteins involved in other pathways were less abundant in acquired resistance biofilm. Targeting the up-regulation of fatty acid biosynthesis, the authors found that a mixture of CTC and triclosan (a fatty-acid-biosynthesis inhibitor) had a more powerful antimicrobial effect than either one of them alone. This information is highly relevant in the fight against this pathogen when forming biofilms, which are always a challenge in seafood farming and processing [105].

Two different quantitative proteomic studies, using dimethyl labeling and label-free methods, performed on the same year as the previous work, were conducted to examine the differential regulation of proteins in response to several doses of oxytetracycline (OXY) in *A. hydrophila* [103].

The results showed an increase in translation-related proteins, although the amount of many central metabolic-related proteins decreased upon OXY treatment and, also, antibiotic sensitivity seemed to be significantly inhibited by numerous external metabolites when they were compounded with OXY antibiotics.

In 2018, Li et al. published a quantitative proteomics experiment based on iTRAQ methodology, to compare proteins differentially regulated in CTC-resistant *A. hydrophila* and in control strains [106]. The majority of the detected differentially regulated proteins were involved in key energy biosynthesis pathways, such as metabolic and catabolic processes, transportation and signal transduction. Chemotaxis-related proteins were downregulated in CTC-resistant strains, but exogenous metabolite addition increased bacterial susceptibility in *A. hydrophila*. In addition, Elbehiry and colleagues described the application of MALDI-TOF MS for the discrimination of the *Aeromonas* genus from meat and water samples, with a spotlight on the antimicrobial resistance of *A. hydrophila* [107].

Recently, another proteomic study using 2-DE and MALDI-TOF/TOF analysis was conducted by Zhu and coworkers with multidrug resistant (MDR) and sensitive *A. hydrophila* strains to find differences in the regulation of proteins [108]. The work showed that, in the sensitive strains, proteins engaged in glycolysis/gluconeogenesis and antibiotic biosynthesis were up regulated in the MDR strain, while those involved in biosynthesis of secondary metabolites, cationic antimicrobial peptide resistance, metabolic processes related to carbon regulation and bacterial metabolism were downregulated. Other proteomics approaches have been used to obtain knowledge about antibiotic resistance in other pathogenic genera, such as *Edwardsiella*, a gram-negative microorganism that also generates hemorrhagic septicemia in a broad number of cultivated fish species, including yellowtail carp and eels. As in the case of *Aeromonas*, the antibiotic resistance status of *Edwardsiella tarda* is of high relevance for seafood safety, particularly when the bacterium is forming biofilms. Sun and coworkers, used iTRAQ-based quantitative proteomics and high-resolution LC-MS/MS to analyze the differential protein regulation of *E. tarda* in response to OXY stress in biofilms [109]. Their work showed a total of 281 modified proteins, 193 of which were downregulated and 88 upregulated. As *A. hydrophila* in biofilms, many ribosomal proteins were upregulated in response to the stress in *E. tarda*, while treatment with OXY increased the amount of Uvr C, a member of the UvrABC system that plays an important role in multiple antibiotic resistance processes.

iTRAQ and LC-MS/MS were used in conjunction to evidence the differential proteome of the ampicillin-resistant LTB4 (LTB4-RAMP) strain of the gram-negative facultative aerobic bacteria *Edwardsiella piscicida* and showed that a depressed P cycle seemed to be a characteristic of the differential proteome in the LTB4-RAMP [110] strain, leading the authors to conclude that the depressed P cycle caused the ampicillin resistance in *E. piscicida*.

The above research works in aquatic organisms, using iTRAQ technologies, seem to indicate that the acquisition of antibiotic resistance involves chemotaxis, energy metabolism, biofilm characteristics and external membrane proteins, as well as networks of proteins associated to antibiotic resistance.

Finally, "reprogramming proteomics" needs to be developed in a general manner to revert an antibiotic-resistance proteome to an antibiotic-sensitive proteome for the control of antibiotic-resistant pathogens [104].

4. Concluding Remarks and Future Directions

As presented in this review, proteomic approaches help to characterize some of the principal issues associated to farming conditions and to address some of the main challenges in aquaculture, such as dietary management, fish welfare, stress responses, food safety and antibiotic resistance.

Proteomics helps to elucidate how dietary management in the aquaculture sector influences the production, growth, immunity and wellness/welfare of living aquaculture organisms and assists in the selection of optimal diets. A large number of publications have shown that the composition of the feed influences the fish muscle nutritional value and its proteome. Efforts have been made to discover protein markers for such quality traits. In addition, various proteomics investigations have been published on the identification of robust protein signatures for fish chronic stress. From this perspective, amelioration of aquaculture conditions to reduce chronic stress during farming and maintain fish welfare is one of the principal issues that can be addressed with discovery proteomics. Additionally, innovative fast targeted proteomics workflows have demonstrated the rapid detection of fish allergens, parasites and microorganisms in aquaculture. The characterization of species-specific peptides by MS/MS-based proteomics and their monitoring by targeted proteomics demonstrated the adequacy of these approaches for food safety control, enabling the differential detection of several hazards in the aquaculture sector. In this way, the utilization of rapid sample preparation methods, combined with sensitive and accurate MS for both the discovery and targeting of fish quality and safety biomarkers, may enhance quality control and safety in aquaculture. Moreover, proteomics offers a more holistic point of view on the molecular mechanisms of antibiotic resistance in the aquaculture sector and it can be directly linked to the metagenomic/metaproteomic approaches that are being applied to the study of a new concept known as the resistome, a current challenge of high relevance that needs an effective and rapid response and that may be elucidated through proteomics techniques.

As the proteins are considered the principal functional macromolecules in all biological systems, we consider that proteomics strategies and their associated techniques can offer several advantages compared with other methodologies for the study of the impact of farming conditions on food quality and safety in aquaculture products. This is the case primarily because, with those methodologies, it is possible to identify and directly quantify protein/peptide signatures without the necessity of inferring conclusions based on other approaches such as genomics tools. Secondly, the benefits of proteomic analysis may be adapted for fish products with short shelf-life. Finally, the current advances in proteomic methodologies allow for the implementation of precise methods that may be useful for routine control test with a potentially lower cost and in a relatively short estimated time (<30 min).

Lastly, the development and practical implementation of new advances based on protein arrays, microfluidics, and biosensors to the aquaculture sector offers a promising research area in which the results of proteomic studies can be established for the routine control test and diagnosis of fish products. We also assume that the digitalization of these new devices may be relevant to the aquaculture industry and control authorities in the next several years and may supply rapid monitoring information to effectively drive decision enforcing by the industry and authorities.

Author Contributions: All authors listed have made a substantial, direct, and intellectual contribution to the work, and have approved it for publication. All authors have read and agreed to the published version of the manuscript.

Funding: This research was funded by GAIN-Xunta de Galicia Project (IN607D 2017/01) and the Agencia Estatal de Investigación (AEI) of Spain and the European Regional Development Fund through project CTM2017-84763-C3-1-R. M.C. is supported by the Ramón y Cajal Contract (RYC-2016-20419, Ministry of Science, Innovation and Universities of Spain).

Conflicts of Interest: The authors declare no conflict of interest.

References

1. Food and Agriculture Organization of the United Nations (FAO). *The State of World Fisheries and Aquaculture 2018: Meeting the Sustainable Development Goals*; Food and Agriculture Organization of the United Nations: Rome, Italy, 2008.
2. Chintagari, S.; Hazard, N.; Edwards, H.G.; Jadeja, R.; Janes, M. Risks associated with fish and seafood. *Microbiol. Spectr.* **2017**, *5*, 1–16. [CrossRef]
3. D'Alessandro, A.; Zolla, L. We are what we eat: Food safety and proteomics. *J. Proteome Res.* **2012**, *11*, 26–36. [CrossRef]
4. Rodrigues, P.M.; Silva, T.S.; Dias, J.; Jessen, F. Proteomics in aquaculture: Applications and trends. *J. Proteom.* **2012**, *75*, 4325–4345. [CrossRef]
5. Rodrigues, P.M.; Campos, A.; Kuruvilla, J.; Schrama, D.; Cristobal, S. Proteomics in Aquaculture: Quality and safety. In *Proteomics in Food Science, from Farm to Fork*; Colgrave, M.L., Ed.; Elsevier: London, UK, 2017; pp. 279–290.

6. Pandey, A.; Mann, M. Proteomics to study genes and genomes. *Nature* **2000**, *405*, 837–846. [CrossRef] [PubMed]
7. Aebersold, R.; Mann, M. Mass-spectrometric exploration of proteome structure and function. *Nature* **2016**, *537*, 347–355. [CrossRef] [PubMed]
8. Carrera, M.; Cañas, B.; Gallardo, J.M. Proteomics for the assessment of quality and safety of fishery products. *Food Res. Int.* **2013**, *54*, 972–979. [CrossRef]
9. Piñeiro, C.; Carrera, M.; Cañas, B.; Lekube, X.; Martínez, I. Proteomics and food analysis: Principles, techniques and applications. In *Handbook of Food Analysis-Two Volume Set*; CRC Press: Boca Raton, FL, USA, 2015; pp. 393–416.
10. Holton, T.A.; Vijayakumar, V.; Khaldi, N. Bioinformatics: Current perspectives and future directions for food and nutritional research facilitated by a Food-Wiki database. *Trends Food Sci. Technol.* **2013**, *34*, 5–17. [CrossRef]
11. Gallardo, J.M.; Carrera, M.; Ortea, I. Proteomics in food science. In *Foodomics: Advanced Mass Spectrometry in Modern Food Science and Nutrition*; Cifuentes, A., Ed.; JohnWiley & Sons Inc.: Hoboken, NJ, USA, 2013; pp. 125–165.
12. Carrera, M.; Mateos, J.; Gallardo, J.M. Data treatment in food proteomics. In *Reference Module in Food Science*; Elsevier: London, UK, 2019. [CrossRef]
13. Rabilloud, T.; Lelong, C. Two-dimensional gel electrophoresis in proteomics: A tutorial. *J. Proteom.* **2011**, *74*, 1829–1841. [CrossRef]
14. Carrera, M.; Cañas, B.; Piñeiro, C.; Vázquez, J.; Gallardo, J.M. De novo mass spectrometry sequencing and characterization of species-specific peptides from nucleoside diphosphate kinase B for the classification of commercial fish species belonging to the family Merlucciidae. *J. Proteome Res.* **2007**, *6*, 3070–3080. [CrossRef]
15. Wolters, D.A.; Washburn, M.P.; Yates, J.R., 3rd. An automated multidimensional protein identification technology for shotgun proteomics. *Anal. Chem.* **2001**, *73*, 5683–5690. [CrossRef]
16. Carrera, M.; Ezquerra-Brauer, J.M.; Aubourg, S.P. Characterization of the jumbo squid (*Dosidicus gigas*) skin by-product by shotgun proteomics and protein-based bioinformatics. *Mar. Drugs* **2019**, *18*, 31. [CrossRef] [PubMed]
17. Zhang, Y.; Fonslow, B.R.; Shan, B.; Baek, M.C.; Yates, J.R., 3rd. Protein analysis by shotgun/bottom-up proteomics. *Chem. Rev.* **2013**, *113*, 2343–2394. [CrossRef]
18. Perkins, D.N.; Pappin, D.J.C.; Creasy, D.M.; Cottrell, J.S. Probability-based protein identification by searching sequence databases using mass spectrometry data. *Electrophoresis* **1999**, *20*, 3551–3567. [CrossRef]
19. Eng, J.K.; McCormack, A.L.; Yates, J.R.I.I.I. An approach to correlate tandem mass spectral data of peptides with amino acid sequences in a protein database. *J. Am. Soc. Mass Spectrom.* **1994**, *5*, 976–989. [CrossRef]
20. Kall, L.; Canterbury, J.D.; Weston, J.; Noble, W.S.; MacCoss, M.J. Semi-supervised learning for peptide identification from shotgun proteomics datasets. *Nat. Methods* **2007**, *4*, 923–925. [CrossRef] [PubMed]
21. Shevchenko, A.; Wilm, M.; Mann, M. Peptide sequencing by mass spectrometry for homology searches and cloning of genes. *J. Protein Chem.* **1997**, *16*, 481–490. [CrossRef]
22. Ma, B.; Zhang, K.; Hendrie, C.; Liang, C.; Li, M.; Doherty-Kirby, A.; Lajoie, G. PEAKS: Powerful software for peptide de novo sequencing by tandem mass spectrometry. *Rapid Commun. Mass Spectrom.* **2003**, *17*, 2337–2342. [CrossRef]
23. Scigelova, M.; Maroto, F.; Dufresne, C.; Vázquez, J. High Throughput de novo Sequencing. 2007. Available online: http://www.thermo.com/ (accessed on 23 July 2020).
24. Carrera, M.; Cañas, B.; Vázquez, J.; Gallardo, J.M. Extensive de novo sequencing of new parvalbumin isoforms using a novel combination of bottom-up proteomics, accurate molecular mass measurement by FTICR-MS, and selected MS/MS ion monitoring. *J. Proteome Res.* **2010**, *9*, 4393–4406. [CrossRef]
25. Ortea, I.; Cañas, B.; Gallardo, J.M. Mass spectrometry characterization of species-specific peptides from arginine kinase for the identification of commercially relevant shrimp species. *J. Proteome Res.* **2009**, *8*, 5356–5362. [CrossRef]
26. Ong, S.E.; Blagoev, B.; Kratchmarova, I.; Kristensen, D.B.; Steen, H.; Pandey, A.; Mann, M. Stable isotope labelling by amino acids in cell culture, SILAC, as a simple and accurate approach to expression proteomics. *Mol. Cell. Proteom.* **2002**, *1*, 376–386. [CrossRef]

27. Mateos, J.; Landeira-Abia, A.; Fafián-Labora, J.A.; Fernández-Pernas, P.; Lesende-Rodríguez, I.; Fernández-Puente, P.; Fernández-Moreno, M.; Delmiro, A.; Martín, M.A.; Blanco, F.J.; et al. iTRAQ-based analysis of progerin expression reveals mitocondrial dysfunction, reactive oxygen species accumulation and altered proteostasis. *Stem Cell Res. Ther.* **2015**, *6*, 119. [CrossRef]
28. Robotti, E.; Marengo, E. 2D-DIGE and fluorescence image analysis. *Methods Mol. Biol.* **2018**, *1664*, 25–39. [CrossRef] [PubMed]
29. Stryiński, R.; Mateos, J.; Pascual, S.; González, A.F.; Gallardo, J.M.; Łopieńska-Biernat, E.; Medina, I.; Carrera, M. Proteome profiling of L3 and L4 *Anisakis simplex* development stages by TMT-based quantitative proteomics. *J. Proteom.* **2019**, *201*, 1–11. [CrossRef]
30. López-Ferrer, D.; Ramos-Fernández, A.; Martínez-Bartolomé, S.; García-Ruiz, P.; Vázquez, J. Quantitative proteomics using $^{16}O/^{18}O$ labeling and linear ion trap mass spectrometry. *Proteomics* **2006**, *6* (Suppl. S1), S4–S11. [CrossRef]
31. Mueller, L.N.; Rinner, O.; Schmidt, A.; Letarte, S.; Bodenmiller, B.; Brusniak, M.Y.; Vitek, O.; Aebersold, R.; Müller, M. SuperHirn—A novel tool for high resolution LC-MS-based peptide/protein profiling. *Proteomics* **2007**, *7*, 3470–3480. [CrossRef] [PubMed]
32. Borràs, E.; Sabidó, E. What is targeted proteomics? A concise revision of targeted acquisition and targeted data analysis in mass spectrometry. *Proteomics* **2017**, *17*, 17–18. [CrossRef]
33. Aebersold, R.; Bensimon, A.; Collins, B.C.; Ludwig, C.; Sabido, E. Applications and developments in targeted proteomics: From SRM to DIA/SWATH. *Proteomics* **2016**, *16*, 2065–2067. [CrossRef] [PubMed]
34. Lange, V.; Picotti, P.; Domon, B.; Aebersold, R. Selected reaction monitoring for quantitative proteomics: A tutorial. *Mol. Syst. Biol.* **2008**, *4*, 1–14. [CrossRef] [PubMed]
35. Jorge, I.; Casas, E.M.; Villar, M.; Ortega-Pérez, I.; López-Ferrer, D.; Martínez-Ruiz, A.; Carrera, M.; Marina, A.; Martínez, P.; Serrano, H.; et al. High-sensitivity analysis of specific peptides in complex samples by selected MS/MS ion monitoring and linear ion trap mass spectrometry: Application to biological studies. *J. Mass Spectrom.* **2007**, *42*, 1391–1403. [CrossRef]
36. Carrera, M.; Cañas, B.; López-Ferrer, D.; Piñeiro, C.; Vázquez, J.; Gallardo, J.M. Fast monitoring of species-specific peptide biomarkers using high-intensity-focused-ultrasound-assisted tryptic digestion and selected MS/MS ion monitoring. *Anal. Chem.* **2011**, *83*, 5688–5695. [CrossRef]
37. Carrera, M.; Gallardo, J.M.; Pascual, S.; González, A.F.; Medina, I. Protein biomarker discovery and fast monitoring for the identification and detection of Anisakids by parallel reaction monitoring (PRM) mass spectrometry. *J. Proteom.* **2016**, *142*, 130–137. [CrossRef] [PubMed]
38. Gillet, L.C.; Navarro, P.; Tate, S.; Röst, H.; Selevsek, N.; Reiter, L.; Bonner, R.; Aebersold, R. Targeted data extraction of the MS/MS spectra generated by data-independent acquisition: A new concept for consistent and accurate proteome analysis. *Mol. Cell. Proteom.* **2012**, *11*, 016717. [CrossRef] [PubMed]
39. Beynon, R.J.; Doherty, M.K.; Pratt, J.M.; Gaskell, S.J. Multiplexed absolute quantification in proteomics using artificial QCAT proteins of concatenated signature peptides. *Nat. Methods* **2005**, *2*, 587–589. [CrossRef] [PubMed]
40. Röst, H.; Malmström, L.; Aebersold, R. A computational tool to detect and avoid redundancy in selected reaction monitoring. *Mol. Cell. Proteom.* **2012**, *11*, 540–549. [CrossRef]
41. Bereman, M.S.; MacLean, B.; Tomazela, D.M.; Liebler, D.C.; MacCoss, M.J. The development of selected reaction monitoring methods for targeted proteomics via empirical refinement. *Proteomics* **2012**, *12*, 1134–1141. [CrossRef] [PubMed]
42. Carpene, E.; Martin, B.; Dalla Libera, L. Biochemical differences in lateral muscle of wild and farmed gilthead sea bream (series *Sparus aurata* L.). *Fish Physiol. Biochem.* **1998**, *19*, 229–238. [CrossRef]
43. Martinez, I.; Standal, I.B.; Aursand, M.; Yamashita, Y.; Yamashita, M. Analytical Methods to differentiate farmed from wild seafood. In *Handbook of Seafood and Seafood Products Analysis*; Nollet, L., Toldrá, F., Eds.; CRC Press: Boca Raton, FL, USA, 2010; pp. 215–232.
44. Chiozzi, R.Z.; Capriotti, A.L.; Cavalieri, C.; La Barbera, G.; Montone, C.M.; Piovesana, S.; Lagana, A. Label-Free shotgun proteomics approach to characterize muscle tissue from farmed and wild European sea bass (*Dicentrarchus labrax*). *Food Anal. Method.* **2018**, *11*, 292–301. [CrossRef]

45. Torstensen, B.E.; Espe, M.; Sanden, M.; Stubhaug, I.; Waagbø, R.; Hemre, G.I.; Fontanillas, F.; Nordgarden, U.; Hevrøy, E.M.; Olsvik, P.; et al. Novel production of Atlantic salmon (*Salmo salar*) protein based on combined replacement of fish meal and fish oil with plant meal and vegetable oil blends. *Aquaculture* **2008**, *285*, 193–200. [CrossRef]
46. Dalmo, R.A.; Bøgwald, J. ß-glucans as conductors of immune symphonies. *Fish Shellfish Immunol.* **2008**, *25*, 384–396. [CrossRef]
47. Ghaedi, G.; Keyvanshokooh, S.; Mohammadi Azarm, H.; Akhlaghi, M. Proteomic analysis of muscle tissue from rainbow trout (*Oncorhynchus mykiss*) fed dietary β-glucan. *Iran. J. Vet. Res.* **2016**, *17*, 184–189.
48. Martin, S.A.M.; Vilhelmsson, O.; Médale, F.; Watt, P.; Kaushik, S.; Houlihan, D.F. Proteomic sensitivity to dietary manipulations in rainbow trout. *Biochim. Biophys. Acta* **2003**, *1651*, 17–29. [CrossRef]
49. Estruch, G.; Martínez-Llorens, S.; Tomás-Vidal, A.; Monge-Ortiz, R.; Jover-Cerdá, M.; Brown, P.B.; Peñaranda, D.S. Impact of high dietary plant protein with or without marine ingredients in gut musosa proteome of gilthead seabream (*Sparus aurata*, L.). *J. Proteom.* **2020**, *216*, 103672. [CrossRef] [PubMed]
50. Nasopoulou, C.; Zabetakis, I. Benefits of fish oil replacement by plant originated oils in compounded fish feeds. A review. *LWT* **2012**, *47*, 217–224. [CrossRef]
51. Morais, S.; Silva, T.; Cordeiro, O.; Rodrigues, P.; Guy, D.R.; Bron, J.E.; Taggart, J.B.; Bell, J.G.; Tocher, D.R. Effects of genotype and dietary fish oil replacement with vegetable oil on the intestinal transcriptome and proteome of Atlantic salmon (*Salmo salar*). *BMC Genom.* **2012**, *13*, 448. [CrossRef]
52. Monti, G.; De Napoli, L.; Mainolfi, P.; Barone, R.; Guida, M.; Marino, G.; Amoresano, A. Monitoring food quality by microfluidic electrophoresis, gas chromatography, and mass spectrometry techniques: Effects of aquaculture on the sea bass (*Dicentrarchus labrax*). *Anal. Chem.* **2005**, *77*, 2587–2594. [CrossRef]
53. Belghit, I.; Lock, E.J.; Fumière, O.; Lecrenier, M.C.; Renard, P.; Dieu, M.; Berntssen, M.H.G.; Palmblad, M.; Rasinger, J.D. Species-specific discrimination of insect meals for aquafeeds by direct comparison of tandem mass spectra. *Animals* **2019**, *9*, 222. [CrossRef]
54. Marco-Ramell, A.; de Almeida, A.M.; Cristobal, S.; Rodrigues, P.; Roncada, P.; Bassols, A. Proteomics and the search for welfare and stress biomarkers in animal production in the one-health context. *Mol. Biosyst.* **2016**, *12*, 2024–2035. [CrossRef]
55. Raposo de Magalhães, C.; Schrama, D.; Farinha, A.P.; Revets, D.; Kuehn, A.; Planchon, S.; Rodrigues, P.M.; Marco Cerqueira, M.A. Protein changes as robust signatures of fish chronic stress: A proteomics approach in fish welfare research. *BMC Genom.* **2020**, *21*, 309. [CrossRef]
56. Mommsen, T.P.; Vijayan, M.M.; Moon, T.W. Cortisol in teleosts: Dynamics, mechanisms of action, and metabolic regulation. *Rev. Fish Biol. Fish.* **1999**, *9*, 211–268. [CrossRef]
57. Torres-Velarde, J.; Llera-Herrera, R.; García-Gasca, T.; García-Gasca, A. Mechanisms of stress-related muscle atrophy in fish: An ex vivo approach. *Mech. Dev.* **2018**, *154*, 162–169. [CrossRef]
58. Christiansen, J.S.; Ringø, E.; Jobling, M. Effects of sustained exercise on growth and body composition of first-feeding fry of Arctic charr, *Salvelinus alpinus* (L.). *Aquaculture* **1989**, *79*, 329–335. [CrossRef]
59. Christiansen, J.S.; Jobling, M. The behaviour and the relationship between food intake and growth of juvenile Arctic charr, *Salvelinus alpinus* L. subjected to sustained exercise. *Can. J. Zool.* **1990**, *68*, 2185–2191. [CrossRef]
60. Jobling, M.; Baardvik, B.M.; Christiansen, J.S.; Jørgensen, E.H. The effects of prolonged exercise training on growth performance and production parameters in fish. *Aquac. Int.* **1993**, *1*, 95–111. [CrossRef]
61. Eguiraun, H.; Casquero, O.; Sørensen, A.J.; Martinez, I. Reducing the number of individuals to monitor shoaling fish systems—Application of the Shannon entropy to construct a biological warning system model. *Front. Physiol.* **2018**, *9*, 493. [CrossRef]
62. Christiansen, J.S.; Martinez, I.; Jobling, M.; Amin, A. Rapid somatic growth and muscle damage in a salmonid fish. *Basic Appl. Myol.* **1992**, *2*, 235–239.
63. Stickland, N.C. Growth and development of muscle fibres in the rainbow trout (*Salmo gairdneri*). *J. Anat.* **1983**, *137*, 323–333. [PubMed]
64. Rossi, G.; Messina, G. Comparative myogenesis in teleosts and mammals. *Cell. Mol. Life Sci.* **2014**, *71*, 3081–3099. [CrossRef] [PubMed]
65. Nemova, N.N.; Lysenko, L.A.; Kantserova, N.P. Degradation of skeletal muscle protein during growth and development of salmonid fish. *Russ. J. Dev. Biol.* **2016**, *47*, 161–172. [CrossRef]

66. Vélez, E.J.; Lutfi, E.; Azizi, S.; Perelló, M.; Salmerón, C.; Riera-Codina, M.; Ibarz, A.; Fernández-Borràs, J.; Blasco, J.; Capilla, E.; et al. Understanding fish muscle growth regulation to optimize aquaculture production. *Aquaculture* **2017**, *467*, 28–40. [CrossRef]
67. Stockdale, F.E. Myogenic cell lineages. *Dev. Biol.* **1992**, *154*, 284–298. [CrossRef]
68. Bigard, A.X.; Janmot, C.; Sanchez, H.; Serrurier, B.; Pollet, S.; d'Albis, A. Changes in myosin heavy chain profile of mature regenerated muscle with endurance training in rat. *Acta Physiol. Scand.* **1999**, *165*, 185–192. [CrossRef] [PubMed]
69. Addis, M.F.; Cappuccinelli, R.; Tedde, V.; Pagnozzi, D.; Porcu, M.C.; Bonaglini, E.; Roggio, T.; Uzzau, S. Proteomic analysis of muscle tissue from gilthead sea bream (*Sparus aurata*, L.) farmed in offshore floating cages. *Aquaculture* **2010**, *309*, 245–252. [CrossRef]
70. Silva, T.S.; Cordeiro, O.D.; Matos, E.D.; Wulff, T.; Dias, J.P.; Jessen, F.; Rodrigues, P.M. Effects of preslaughter stress levels on the post-mortem sarcoplasmic proteomic profile of gilthead seabream muscle. *J. Agric. Food Chem.* **2012**, *60*, 9443–9453. [CrossRef] [PubMed]
71. Xu, Z.N.; Zheng, G.D.; Wu, C.B.; Jiang, X.Y.; Zou, S.M. Identification of proteins differentially expressed in the gills of grass carp (*Ctenopharyngodon idella*) after hypoxic stress by two-dimensional gel electrophoresis analysis. *Fish Physiol. Biochem.* **2019**, *45*, 743–752. [CrossRef]
72. Timmins-Schiffman, E.B.; Crandall, G.A.; Vadopalas, B.; Riffle, M.E.; Nunn, B.L.; Roberts, S.B. Integrating discovery-driven proteomics and selected reaction monitoring to develop a noninvasive assay for geoduck reproductive maturation. *J. Proteome Res.* **2017**, *16*, 3298–3309. [CrossRef]
73. Spencer, L.H.; Horwith, M.; Lowe, A.T.; Venkataraman, Y.R.; Timmins-Schiffman, E.; Nunn, B.L.; Roberts, S.B. Pacific geoduck (*Panopea generosa*) resilience to natural pH variation. *Comp. Biochem. Physiol. Part D Genomics Proteom.* **2019**, *30*, 91–101. [CrossRef]
74. Freitas, J.; Vaz-Pires, P.; Câmara, J.S. From aquaculture production to consumption: Freshness, safety, traceability and authentication, the four pillars of quality. *Aquaculture* **2020**, *518*, 734857. [CrossRef]
75. Teklemariam, A.D.; Tessema, F.; Abayneh, T. Review on evaluation of safety of fish and fish products. *Int. J. Fish Aquat. Stud.* **2015**, *3*, 111–117.
76. Huss, H.H. Assurance of Seafood Quality. In *FAO Fishery Technical Paper No. 334*; FAO: Rome, Italy, 1994; p. 169.
77. Jahncke, M.L.; Schwarz, M.H. Public, animal and environmental aquaculture health issues in industrialized countries. In *Public, Animal and Environmental Aquaculture Health Issues*; Jahncke, M., Garrett, E.S., Reilly, A., Martin, R.E., Cole, E., Eds.; John Wiley & Sons, Inc.: New York, NY, USA, 2002; pp. 67–102.
78. Carrera, M.; Böhme, K.; Gallardo, J.M.; Barros-Velázquez, J.; Cañas, B.; Calo-Mata, P. Characterization of foodborne strains of *Staphylococcus aureus* by shotgun proteomics: Functional networks, virulence factors and species-specific peptide biomarkers. *Front. Microbiol.* **2017**, *8*, 2458. [CrossRef]
79. Hazen, T.H.; Martinez, R.J.; Chen, Y.F.; Lafon, P.C.; Garrett, N.M.; Parsons, M.B.; Bopp, C.A.; Sullards, M.C.; Sobecky, P.A. Rapid identification of *Vibrio parahaemolyticus* by whole-cell matrix-assisted laser desorption ionization-time of flight mass spectrometry. *Appl. Environ. Microbiol.* **2009**, *75*, 6745–6756. [CrossRef]
80. Kazazić, S.P.; Topić Popović, N.; Strunjak-Perović, I.; Babić, S.; Florio, D.; Fioravanti, M.; Bojanić, K.; Čož-Rakovac, R. Matrix-assisted laser desorption/ionization time of flight mass spectrometry identification of *Vibrio* (*Listonella*) *anguillarum* isolated from sea bass and sea bream. *PLoS ONE* **2019**, *14*, e0225343. [CrossRef] [PubMed]
81. Li, J.N.; Zhao, Y.T.; Cao, S.L.; Wang, H.; Zhang, J.J. Integrated transcriptomic and proteomic analyses of grass carp intestines after vaccination with a double-targeted DNA vaccine of *Vibrio mimicus*. *Fish Shellfish Immunol.* **2020**, *98*, 641–652. [CrossRef] [PubMed]
82. Böhme, K.; Fernández-No, I.C.; Barros-Velázquez, J.; Gallardo, J.M.; Calo-Mata, P.; Cañas, B. Species differentiation of seafood spoilage and pathogenic gram-negative bacteria by MALDI-TOF mass fingerprinting. *J. Proteome Res.* **2010**, *9*, 3169–3183. [CrossRef]
83. Piamsomboon, P.; Jaresitthikunchai, J.; Hung, T.Q.; Roytrakul, S.; Wongtavatchai, J. Identification of bacterial pathogens in cultured fish with a custom peptide database constructed by matrix-assisted laser desorption/ionization time-of-flight mass spectrometry (MALDI-TOF MS). *BMC Vet. Res.* **2020**, *16*, 52. [CrossRef] [PubMed]
84. Saleh, S.; Staes, A.; Deborggraeve, S.; Gevaert, K. Targeted proteomics for studying pathogenic bacteria. *Proteomics* **2019**, *19*, 1–10. [CrossRef] [PubMed]

85. Li, Y.; Yang, B.; Tian, J.; Sun, W.; Wang, G.; Qian, A.; Wang, C.; Shan, X.; Kang, Y. An iTRAQ-based comparative proteomics analysis of the biofilm and planktonic states of *Aeromonas veronii* TH0426. *Int. J. Mol. Sci.* **2020**, *21*, 1450. [CrossRef]
86. Saleh, M.; Kumar, G.; Abdel-Baki, A.S.; Dkhil, M.A.; El-Matbouli, M.; Al-Quraishy, S. Quantitative proteomic profiling of immune responses to *Ichthyophthirius multifiliis* in common carp skin mucus. *Fish Shellfish Immunol.* **2019**, *84*, 834–842. [CrossRef]
87. Sangsuriya, P.; Huang, J.Y.; Chu, Y.F.; Phiwsaiya, K.; Leekitcharoenphon, P.; Meemetta, W.; Senapin, S.; Huang, W.P.; Withyachumnarnkul, B.; Flegel, T.W.; et al. Construction and application of a protein interaction map for white spot syndrome virus (WSSV). *Mol. Cell. Proteom.* **2014**, *13*, 269–282. [CrossRef]
88. Chan, L.L.; Sit, W.H.; Lam, P.K.; Hsieh, D.P.; Hodgkiss, I.J.; Wan, J.M.; Ho, A.Y.; Choi, N.M.; Wang, D.Z.; Dudgeon, D. Identification and characterization of a "biomarker of toxicity" from the proteome of the paralytic shellfish toxin-producing dinoflagellate *Alexandrium tamarense* (Dinophyceae). *Proteomics* **2006**, *6*, 654–666. [CrossRef]
89. Carrera, M.; Garet, E.; Barreiro, A.; Garcés, E.; Pérez, D.; Guisande, C.; González-Fernández, A. Generation of monoclonal antibodies for the specific immunodetection of the toxic dinoflagellate *Alexandrium minutum* Halim from Spanish waters. *Harmful Algae* **2010**, *9*, 272–280. [CrossRef]
90. Hennon, G.M.M.; Dyhrman, S.T. Progress and promise of omics for predicting the impacts of climate change on harmful algal blooms. *Harmful Algae* **2020**, *91*, 101587. [CrossRef] [PubMed]
91. Piñeiro, C.; Cañas, B.; Carrera, M. The role of proteomics in the study of the influence of climate change on seafood products. *Food Res. Int.* **2010**, *43*, 1791–1802. [CrossRef]
92. Carrera, M.; Cañas, B.; Gallardo, J.M. Advanced proteomics and systems biology applied to study food allergy. *Curr. Opin. Food Sci.* **2018**, *22*, 9–16. [CrossRef]
93. Carrera, M.; Cañas, B.; Gallardo, J.M. Rapid direct detection of the major fish allergen, parvalbumin, by selected MS/MS ion monitoring mass spectrometry. *J. Proteom.* **2012**, *75*, 3211–3220. [CrossRef]
94. Abdel Rahman, A.M.; Kamath, S.; Lopata, A.L.; Helleur, R.J. Analysis of the allergenic proteins in black tiger prawn (*Penaeus monodon*) and characterization of the major allergen tropomyosin using mass spectrometry. *Rapid Commun. Mass Spectrom.* **2010**, *24*, 2462–2470. [CrossRef] [PubMed]
95. De Magalhães, C.R.; Schrama, D.; Fonseca, F.; Kuehn, A.; Morisset, M.; Ferreira, S.R.; Gonçalves, A.; Rodrigues, P.M. Effect of EDTA enriched diets on farmed fish allergenicity and muscle quality; a proteomics approach. *Food Chem.* **2020**, *305*, 125508. [CrossRef]
96. Mi, J.; Orbea, A.; Syme, N.; Ahmed, M.; Cajaraville, M.P.; Cristobal, S. Peroxisomal proteomics, a new tool for risk assessment of peroxisome proliferating pollutants in the marine environment. *Proteomics* **2005**, *5*, 3954–3965. [CrossRef]
97. Apraiz, I.; Leoni, G.; Lindenstrand, D.; Persson, J.O.; Cristobal, S. Proteomic analysis of mussels exposed to fresh and weathered Prestige's oil. *J. Proteom. Bioinf.* **2009**, *2*, 255–261. [CrossRef]
98. Zhang, Q.H.; Huang, L.; Zhang, Y.; Ke, C.H.; Huang, H.Q. Proteomic approach for identifying gonad differential proteins in the oyster (*Crassostrea angulata*) following food-chain contamination with $HgCl_2$. *J. Proteom.* **2003**, *94*, 37–53. [CrossRef]
99. Campos, A.; Danielsson, G.; Farinha, A.P.; Kuruvilla, J.; Warholm, P.; Cristobal, S. Shotgun proteomics to unravel marine mussel (*Mytilus edulis*) response to long-term exposure to low salinity and propranolol in a Baltic Sea microcosm. *J. Proteom.* **2016**, *137*, 97–106. [CrossRef]
100. Gouveia, D.; Chaumot, A.; Charnot, A.; Almunia, C.; François, A.; Navarro, L.; Armengaud, J.; Salvador, A.; Geffard, O. Ecotoxico-proteomics for aquatic environmental monitoring: First in situ application of a new proteomics-based multibiomarker assay using caged amphipods. *Environ. Sci. Technol.* **2017**, *51*, 13417–13426. [CrossRef] [PubMed]
101. Barros, D.; Pradhan, A.; Mendes, V.M.; Manadas, B.; Santos, P.M.; Pascoal, C.; Cássio, F. Proteomics and antioxidant enzymes reveal different mechanisms of toxicity induced by ionic and nanoparticulate silver in bacteria. *Environ. Sci. Nano* **2019**, *6*, 1207–1218. [CrossRef]
102. Lulijwa, R.; Rupia, E.J.; Alfaro, A.C. Antibiotic use in aquaculture, policies and regulation, health and environmental risks: A review of the top 15 major producers. *Rev. Aquac.* **2020**, *12*, 640–663. [CrossRef]
103. Yao, Z.; Li, W.; Lin, Y.; Wu, Q.; Yu, F.; Lin, W.; Lin, X. Proteomic analysis reveals that metabolic flows affect the susceptibility of *Aeromonas hydrophila* to antibiotics. *Sci. Rep.* **2016**, *6*, 39413. [CrossRef]

104. Peng, B.; Li, H.; Peng, X. Proteomics approach to understand bacterial antibiotic resistance strategies. *Expert Rev. Proteom.* **2019**, *16*, 829–839. [CrossRef]
105. Li, W.; Yao, Z.; Sun, L.; Hu, W.; Cao, J.; Lin, W.; Lin, X. Proteomics analysis reveals a potential antibiotic cocktail therapy strategy for *Aeromonas hydrophila* infection in biofilm. *J. Proteome Res.* **2016**, *15*, 1810–1820. [CrossRef]
106. Li, W.; Ali, F.; Cai, Q.; Yao, Z.; Sun, L.; Lin, W.; Lin, X. Reprint of: Quantitative proteomic analysis reveals that chemotaxis is involved in chlortetracycline resistance of *Aeromonas hydrophila*. *J. Proteom.* **2018**, *180*, 138–146. [CrossRef]
107. Elbehiry, A.; Marzouk, E.; Abdeen, E.; Al-Dubaib, M.; Alsayeqh, A.; Ibrahem, M.; Hamada, M.; Alenzi, A.; Moussa, I.; Hemeg, H.A. Proteomic characterization and discrimination of *Aeromonas* species recovered from meat and water samples with a spotlight on the antimicrobial resistance of *Aeromonas hydrophila*. *Microbiologyopen* **2019**, *8*, e782. [CrossRef]
108. Zhu, W.; Zhou, S.; Chu, W. Comparative proteomic analysis of sensitive and multi-drug resistant *Aeromonas hydrophila* isolated from diseased fish. *Microb. Pathog.* **2019**, *139*, 103930. [CrossRef]
109. Sun, L.; Chen, H.; Lin, W.; Lin, X. Quantitative proteomic analysis of *Edwardsiella tarda* in response to oxytetracycline stress in biofilm. *J. Proteom.* **2017**, *150*, 141–148. [CrossRef]
110. Su, Y.B.; Kuang, S.F.; Peng, X.X.; Li, H. The depressed P cycle contributes to the acquisition of ampicillin resistance in *Edwardsiella piscicida*. *J. Proteom.* **2020**, *212*, 103562. [CrossRef] [PubMed]

© 2020 by the authors. Licensee MDPI, Basel, Switzerland. This article is an open access article distributed under the terms and conditions of the Creative Commons Attribution (CC BY) license (http://creativecommons.org/licenses/by/4.0/).

MDPI
St. Alban-Anlage 66
4052 Basel
Switzerland
Tel. +41 61 683 77 34
Fax +41 61 302 89 18
www.mdpi.com

Foods Editorial Office
E-mail: foods@mdpi.com
www.mdpi.com/journal/foods

www.ingramcontent.com/pod-product-compliance
Lightning Source LLC
LaVergne TN
LVHW070556100526
838202LV00012B/483